D0526768

Instant**Facts**

Maths

A-Z of **essential facts** and definitions

William Collins' dream of knowledge for all began with the publication of his first book in 1819. A self-educated mill worker, he not only enriched millions of lives, but also founded a flourishing publishing house. Today, staying true to this spirit, Collins books are packed with inspiration, innovation and practical expertise. They place you at the centre of a world of possibility and give you exactly what you need to explore it.

Collins. Do more.

Published by Collins
An imprint of HarperCollins*Publishers*
77–85 Fulham Palace Road
Hammersmith
London
W6 8JB

Browse the complete Collins catalogue at

www.collinseducation.com
© HarperCollins*Publishers* Limited 2005

First published as Collins Gem Basic Facts Maths 1998

10 9 8 7 6 5 4 3 2

ISBN-13 978 0 00 720512 7
ISBN-10 0 00 720512 0

British Library Cataloguing in Publication Data
A catalogue record for this publication is available from the British Library

Every effort has been made to contact the holders if copyright material, but if any have been inadvertently overlooked, the Publishers will be pleased to make the necessary arrangements at the first opportunity.

Edited and Project Managed by Marie Insall
Production by Katie Butler
Design by Sally Boothroyd/Wendi Watson
Printed and bound by Printing Express, Hong Kong

You might also like to visit
www.harpercollins.co.uk
The book lover's website

Introduction

Instant Facts Maths is one of a series of illustrated A–Z subject reference guides of the key terms and concepts used in the most important school subjects. With its alphabetical arrangement, the book is designed for quick reference to explain the meaning of words used in the subject and so is an excellent companion both to course work and during revision.

Bold words in an entry identify key terms which are explained in greater detail in entries of their own; important terms that do not have separate entries are shown in *italic* and are explained in the entry in which they occur.

Other titles in the *Instant Facts* series include:
English
Modern World History
Biology
Science
Physics
Geography
Chemistry

A

abscissa (*pl.* **abscissae**) The x-coordinate of a point referred to a **Cartesian coordinate** system. For example, in the diagram, the two points P and Q have abscissae of 2 and –3 respectively. See also **ordinate**.

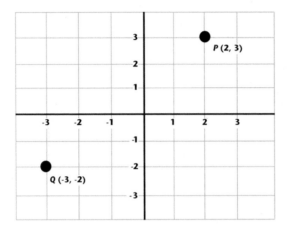

abscissa

absolute value (sometimes called the **modulus**) The size of a real number without regard for **sign**. For example, the absolute value of –7.4 is written as |–7.4| and equals 7.4. This symbol enables mathematicians to write **inequalities** of the type:

$$-3 \leq x \leq 3$$

in the form:

$$|x| \leq 3.$$

acceleration The change in **velocity** of a particle divided by the time taken to make the change. Acceleration is a **vector** quantity describing both the size and direction of a change of velocity. For example, when a car is slowed down its velocity is reduced in the direction of its motion. The diagram shows the velocity V of a slowing car at different times. It is decelerating over the last 10 s of its motion from 15 m/s to 0 m/s. The car's acceleration is thus:

$$(0 \text{ m/s} - 15 \text{ m/s}) \div (15 \text{ s} - 5 \text{ s}) = -15 \text{ m/s} \div 10 \text{ s}$$
$$= -1.5 \text{ m/s}^2,$$

which is also represented by the gradient of its velocity–time graph. The negative sign shows that the change in this case is in fact a deceleration.

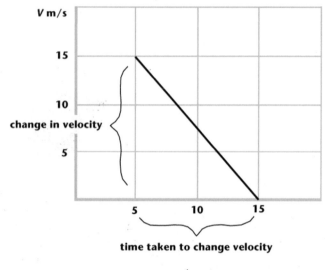

acceleration

accuracy The smallest increment to which a measurement may reasonably be resolved (see **resolve**). In elementary mathematics, the required degree of accuracy of a measurement is often given in an examination question by indicating the number of **decimal places** or **significant** figures required in the solution. If the degree of accuracy is not indicated in the question then one would normally work to one more decimal place or significant figure than has been used in the question (see also **approximation**).

acute angle An **angle** with a value between 0 and 90 **degrees**. See also **obtuse angle**, **right angle**, and **reflex angle**.

addition 1. *Scalar addition.* The **operation** in **arithmetic** that determines the **sum** of two **scalars** and is related to the process of accumulation; for example, 3 + 2 = 5.
2. *Vector addition.* The operation in **vector** analysis that determines the **resultant** of the action of one or more vectors. See **parallelogram rule**.

adjacent Describing **sides**, **angles**, or **faces** that are 'side-by-side'. Two sides of a figure are adjacent if they meet at a point. Two angles are adjacent if they have a common line. Two faces are adjacent if they meet at a common edge.

algebra The method of calculating, using letters to represent numbers and signs to represent the relations between them, making a kind of abstract **arithmetic**. For example, the following algebraic expression:

$$\frac{a}{b} + \frac{c}{d} = \frac{(a \times d) + (c \times b)}{b \times d}$$

may be used to evaluate the **sum** of **fractions** when a, b, c and d take on particular **values**. For $a = 1$, $b = 2$, $c = 1$, $d = 3$:

$$\frac{1}{2} + \frac{1}{3} = \frac{(1 \times 3) + (1 \times 2)}{2 \times 3} = \frac{5}{6}$$

See **Appendix 2**.

algorithm The prescription for a process designed to solve a particular set of problems. Algorithms usually involve a set of simple steps. For instance, the method of calculating the **mean** of a **sample** of **values** is the following algorithm:
(1) Find the sum of the values; call this sum S.
(2) Divide S by the number of values n.
(3) The result, S/n, is the arithmetic mean of the sample.

alternate angles A pair of **angles** that are on opposite sides of a **transversal** intersecting two other lines, with each of the angles having a different line for one of its sides. When these two lines are parallel, the alternate angles are of equal size.

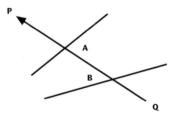

alternate angles: *the angles A and B are alternate angles formed by the line PQ cutting the other two lines*

altitude The height of a figure. In a **triangle**, altitude is the **perpendicular** distance from a **vertex** of the triangle to the opposite side, called the **base**. The area of any triangle can be calculated by using the formula:

area = ½ × base length × altitude.

ambiguous case A mathematical problem in which the given information can lead to two possible solutions or to two possible configurations of a geometric figure.

amplitude The maximum **magnitude** of the **displacement** from the **mean** position of an oscillating **variable**. For the periodic function shown, the amplitude is 2 units (see **periodic**).

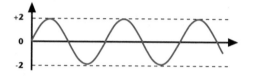

amplitude: *the oscillation is varying between +2 and –2 units from its mean position*

analogue Describing devices that represent measurements as continuous quantities. For example, a meter in a which a pointer moves over a scale is an analogue device. A clock with hands is another example. See also **digital**.

angle The space or shape between two lines; a measure of the **rotation**, about the point of **intersection** of two lines, required to make the lines coincide. The two most common measures of angle are the **degree** and the **radian**. The illustrations summarize some angle theorems:
See also **acute angle**, **alternate angle**, **obtuse angle**, **right angle**, **reflex angle**, and **interior angles**.

angles adjacent on a line
$a + b = 180°$

angles at a point
$a + b + c = 360°$

vertically opposite angles
$a = b$

corresponding angles
$a = b$

interior angles
$a + b = 180°$

alternate angles
$a = b$

angles

annulus A **region** of a **plane** lying between two **concentric circles**. If the circles have **radii** r and R, with $r < R$, then the **area** of the annulus is:

$$\pi R^2 - \pi r^2 \text{ or } \pi(R + r)(R - r).$$

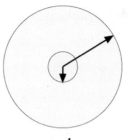

annulus

antilogarithm The **inverse** of a logarithm **function**. For example, since 10 has to be raised by 2 to get 100 (see **logarithm**), one may describe 100 as the antilogarithm to the **base** 10 of 2:

$$\log_{10}(100) = 2$$
$$\text{antilog}_{10}(2) = 100$$
$$\text{or just simply}$$
$$10^2 = 100.$$

apex (*pl.* **apices**) The highest point of a solid or **plane** figure in relation to the **base**.

approximation An inexact result that is accurate enough for some specific purpose.

To approximate a number to d.p. (**decimal** places), decide how many digits are appropriate after the decimal point. Omit all the following digits with the proviso that, if the first digit omitted is 5 or larger, increase the last digit remaining by 1. For example:

75.778 approximated to 1 d.p. is 75.8.

To approximate to s.f. (**significant figures**), count from the first nonzero digit. Zeros may need to be inserted so that the **magnitude** of the number is unchanged. For example:

75.778 approximated to 3 s.f. is 75.8;

75.778 approximated to 1 s.f. is 80.

See also **accuracy** and **estimate**.

arc   Part of the **circumference** of a circle. If the circumference is separated into two parts then the arc that makes the larger **angle** with the centre is the *major arc* and the arc that makes an angle less than 180° at the centre is the *minor arc*. If the circumference is cut exactly in half then two semi-cirular arcs are formed.

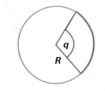

arc: *major and minor arcs of a circle*

are   A unit of land area. The more familiar **hectare** unit (ha) is derived from the are: 1 hectare = 100 ares. Small land areas, such as building plots, are measured in square metres (m^2) with the following equivalences:

1 are = 100 m^2,
1 ha = 10 000 m^2,
100 ha = 1 km2 = 1 000 000 m^2.

There is an appoximate relationship between the **Imperial** (acres) and **metric** (ha) systems of 1 ha = 2.5 acres.

area   The measure of the size of a surface. The **formulae** for the area (A) of some simple shapes are given in the illustrations.

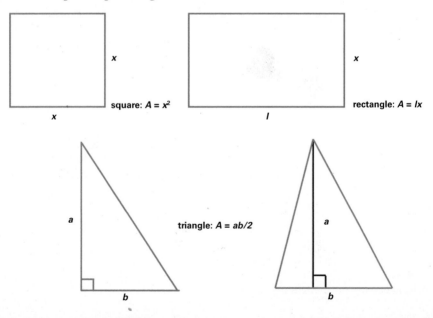

square: $A = x^2$

rectangle: $A = lx$

triangle: $A = ab/2$

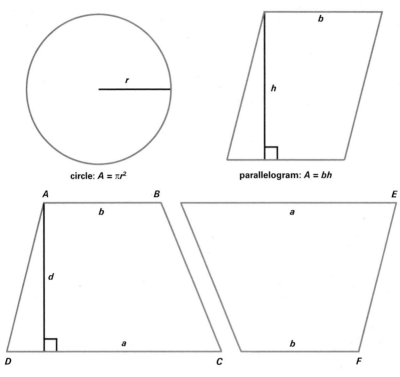

circle: $A = \pi r^2$

parallelogram: $A = bh$

trapezium ABCD = half parallelogram AEFD = $(a + b)d/2$

area under a curve Areas under straight-line graphs are relatively easy to calculate by applying a combination of the formulae for the **areas** of simple shapes. Application of the formula for the area of a **trapezium** is illustrated in the diagram.

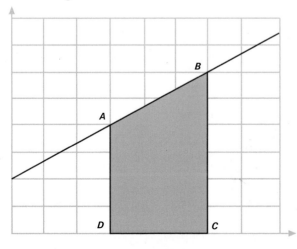

area under a curve: *for the straight-line graph shown, the shaded area is the area of the trapezium ABCD*

If the curve is not a straight line, an approximate value for the area of a shaded region under a curve may be found by a successive application of the above idea. Better estimates of the value may be obtained by partitioning the area into finer and finer trapezia. This is shown in the diagram.

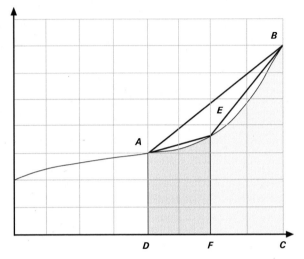

area under a curve: *in the example, the shaded area is approximated by splitting it into trapezia*

The area of the shaded region may be approximated by the area of the trapezium *ABCD*. A better approximation may be obtained by considering the sum of the areas *AEFD* and *EBCF*.

arithmetic The study of numbers and the **operations** associated with their manipulation. Elementary arithmetic is concerned with the simple operations of **addition**, **subtraction**, **multiplication** and **division** and their application to the solutions of problems.

arithmetic mean A single number calculated from a set of n numbers a_1, a_2, .. a_n by adding the numbers together and dividing by n, i.e.:

$$(a_1 + a_2 + ... + a_n)/n.$$

For example, the arithmetic mean of the **set** {5, 7, 1, 8, 4} is:

$$(5 + 7 + 1 + 8 + 4)/5 = 5.$$

The arithmetic mean is often referred to as the **average** of a given set of values.

arithmetic sequence See **sequence**.

asymptote A line on a graph that is approached by a **curve**, but is never reached. For example, the **abscissa** (or x-axis) is an asymptote of the curve $y = 1/x$, i.e. the curve approaches the x-axis for large x-values but never crosses the axis (see "graph sketching" in **Appendix 1**).

average A single number that represents or typifies a collection of values. Three commonly used averages are **mode**, **mean**, and **median**.

average speed See **speed**.

axiom A self-evident principle that does not require proof. For example, 'Things equal to the same thing must be equal to each other', is an axiom attributed to Euclid (see **Euclidean**).

axis (*pl.* **axes**) A fixed line adopted for reference; for example, **coordinate** axes, axes of **symmetry**, etc.

B

balance method A method of solving simple equations of only one unknown. The equation is rearranged so that the unknown is the subject of the formula. See **Appendix 2**.

bar graph A graphical method of representing **data** in terms of **parallel** bars of equal width. The heights or lengths of the bars indicate the frequency of the quantity represented. Only one axis has a scale; the other may represent colour, types of car, numbers of children in a family, etc.

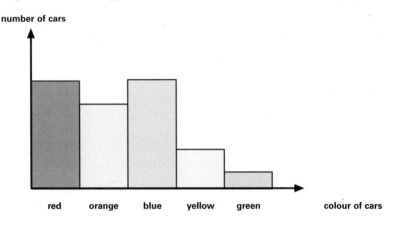

bar graph: *showing the colours of a sample of cars*

bar-line graph A method of representing **data** similar to a **bar graph**, but using single lines instead of bars.

base
1. A number on which a system of numeration is founded. For example, in the **decimal** system any **value** may be represented as a sequence of **digits** based on **powers** of the base 10. In this way the value 472 may be considered to be the **sum**:

$$472 = (4 \times 100) + (7 \times 10) + (2 \times 1) =$$
$$4(10^2) + 7(10^1) + 2(10^0) .$$

2. A number that is raised to a **power**. For example, in the expression 3^4, 4 is the **index** or **exponent** and 3 is called the base. This use of the term base is closely related to its use in the context of logarithms. For example, in logarithms to the base 10, $\log_{10}(47) = 1.6721$ means that $10^{1.6721} = 47$.
3. The lowest side of a **plane** figure or the lowest plane of a **solid** figure.

base units The seven fundamental units used in the international system of units known as SI units (Système International d'Unité). The units are: kilogram (mass), metre (length), second (time), ampere (electrical current), kelvin (temperature increment), mole (amount of substance), and candela (luminous intensity).

bearing The direction or situation of one object with regard to another in navigation or surveying. The bearing of a point A from an observer O is the **angle** between the line AO and the north line through O, measured in a clockwise direction from the north line.

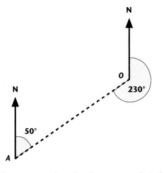

bearing: *in this example, the bearing of A from O is 230°;*
the bearing of O from A is 050°

biased sample A **sample** that is not a representative sample of the **population** under study.

billion A thousand million (10^9). Formerly, the term was taken to mean a million million (10^{12}) in Britain and 10^9 in the USA. The meaning 10^9 is now common to both countries.

bimodal Describing **distributions** of **data** that exhibit two **modes** or peaks in the **frequency**.

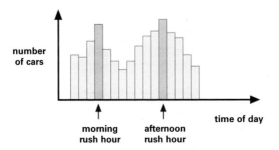

bimodal: *a bimodal distribution of traffic flow*

binary

1. Describing a number system that is founded on **base** 2. Any number in the binary number system may be represented as a sequence of the digits 1 and 0, based on powers of base 2. In this way the number 7 may be written in binary as 111; that is:

$$(1 \times 2^2) + (1 \times 2^1) + (1 \times 2^0) = 1(4) + 1(2) + 1(1) = 7.$$

For example, decimal 29 is given by:

$2^4=16$	$2^3=8$	$2^2=4$	$2^1=2$	$2^0=1$
1	1	1	0	1

or   11101

Decimal 92 is given by:

$2^6=64$	$2^5=32$	$2^4=16$	$2^3=8$	$2^2=4$	$2^1=2$	$2^0=1$
1	0	1	1	1	0	0

or 1011100

2. A rule of combination of two **members** of a set. The four basic **operations** of arithmetic – **addition**, **subtraction**, **multiplication**, and **division** – are all examples of binary operations on the set of **real numbers**.

binomial A **polynomial** involving two variables, for example: $9x + 5y$.

binomial coefficients Coefficients of the terms for the **expansion** of the expression:

$$(x + y)^n, \text{ where } n \in N.$$

These coefficients are given by rows of numbers as shown in the triangular array:

As an example, the fifth row gives the coefficients for the expansion of $(x + y)^4$; that is:

$$(x + y)^4 = \underline{1}x^4 + \underline{4}x^3y + \underline{6}x^2y^2 + \underline{4}xy^3 + \underline{1}y^4.$$

See also **Pascal's triangle**.

bisect To divide into two equal parts. The term is often used with respect to geometrical figures. For example, bisection of an **angle** involves drawing a line through the **vertex** that cuts the angle in half. A point, line or plane that bisects something is called a bisector. Compass constructions of the bisectors of a line and an angle are given in step by step form. (see below)

Bisection of a line:

Step 1: Open out the compass so that the span of the compass is a little more than half the length of the line *AB*. Keep this span on the compass throughout.

Step 2: Place the compass point firstly in *A* and then in *B*, and draw arcs to meet at *C* and *D* as shown above.

Step 3: Draw the line through *C* and *D*. The line *CD* is the bisector of the line AB.

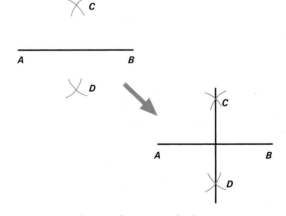

bisect: *bisection of a line*

Bisection of an angle (see illustration on next page):

Step 1: Open out the compass to any reasonable span. This span should be less than the length of either *AB* or *AC*. Keep this span on the compass throughout.

Step 2: Place the compass point in *A* and draw arcs to cut *AB* and *AC* at *X* and *Y*.

Step 3: Place compass firstly on *X* and draw an arc between *B* and *C*, then do the same but this time with the compass point in *Y*. These two arcs cross at a point *Z* as shown.

Step 4: Draw the line through *A* and *Z*. The line *AZ* is the bisector of the angle *BAC*.

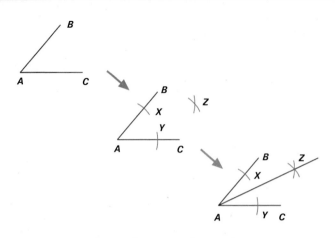

bisect: *bisection of an angle*

boundary The line or surface by which a figure or solid is defined. The boundary of a **polygon** consists of its edges. The boundary of a **circle** is its circumference. The boundary of a **polyhedron** is its surface.

brackets A pair of symbols used to delimit an expression. In mathematics brackets indicate the order in which mathematical operations are to be carried out. For example:

$$8 \times (7 + 5) = 8 \times 12 = 96,$$

$$(8 \times 7) + 5 = 56 + 5 = 61.$$

Elementary **algebra** is often concerned with the introduction or removal of brackets to simplify expressions. For example:

$$6x^2y + 9xy^2 = 3xy(2x + 3y).$$

C

calculate To perform a mathematical operation, often numerical, to obtain a desired result.

calculus A branch of mathematics that deals with the behaviour of **functions**. There are two main types of calculus:
(1) *Differential calculus* deals with how functions change over various values of **parameters**. Using differential calculus one can calculate **turning points** of functions and find their **gradient** functions. One may also use differential calculus to make **approximations** to functions especially for very small incremental changes.
(2) *Integral calculus* deals with calculating **areas** and **volumes** of shapes delimited by functions or parts of functions. See Calculus in **Appendix 1**.

cancellation The manipulation of a **fraction** to leave an **equivalent fraction** by dividing the **numerator** and **denominator** by a common **factor**. To multiply fractions take the following steps:
Step 1: write any whole or mixed numbers as **improper fractions**.
Step 2: make all possible cancellations.
Step 3: multiply the numerators; multiply the denominators.
To divide by a fraction, multiply by the **reciprocal**:
Step 1: find the reciprocal of the fraction; i.e. turn the fraction 'upside down'.
Step 2: follow the steps for multiplying fractions from step 2 above.

$$2^2/_3 \times 2^1/_2 = {}^4\cancel{8}/_3 \times {}^5/\cancel{2}_1$$
$$= {}^{4 \times 5}/_3 = {}^{20}/_3 = 6^2/_3.$$

$$2^2/_5 \div 2^1/_{10} = {}^{12}/_{\cancel{5}} \div {}^{21}/_{10}$$
$$= {}^{4}\cancel{12}/\cancel{5}_1 \times {}^{2}{}^{10}\cancel{/\cancel{21}}_7$$
$$= {}^8/_7 = 1^1/_7.$$

cancellation: *examples in manipulating fractions*

capacity See **volume**.

cardinal number A number that indicates the number of elements in a **set**. For example, the cardinal number of the set of members of a soccer team is 11. The cardinal number 11 is the property shared by all sets containing eleven members.

Cartesian coordinates A system of coordinates in which the position of a **point** is determined by its distances from the **perpendicular** reference **axes**. On a **plane** the distances of a point from the two axes are given as an ordered pair of **real numbers** (x, y). In three-dimensional space, three mutually perpendicular axes may be used as reference lines. The distances of a point from these three axes are given as an ordered triple of real numbers (x, y, z).

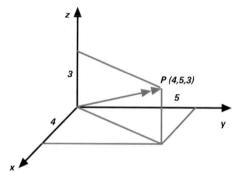

Cartesian coordinates: *coordinates in three dimensions*

Celsius scale A temperature scale. The Celsius scale has the freezing and boiling points of water at 0° and 100°, respectively. There are therefore one hundred graduations between these two fixed points; it is for this reason that the Celsius scale is often called the *centigrade scale*. See also **Fahrenheit**.

centi- A prefix with symbol 'c' meaning 'one-hundredth of'. For example, a centimetre (1 cm) is one-hundredth of a metre.

centigrade scale See **Celsius scale**.

centimetre A **unit** of length: 1 centimetre = 1 hundredth of a metre. See also **unit conversions**.

centre A point inside a figure that is equal distances from all points on the boundary of the figure.

centre of enlargement The point from which an **enlargement** of a plane figure may be thought to originate. The diagram shows how a centre of enlargement can be found. The steps needed to find the centre of enlargement of the construction are as follows:
Step 1: join A' and A.
Step 2: join C' and C.

Step 3: extend the lines $A'A$ and $C'C$. The point P, at which these lines meet, is the centre of enlargement. See also **enlargement** and **scale factor**.

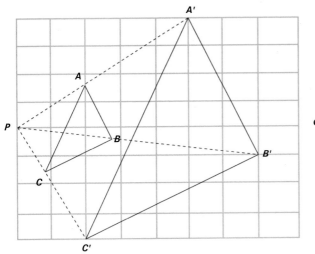

centre of enlargement

centre of symmetry A **point** about which a geometrical configuration may be said to be symmetric. A geometric configuration has a centre of **symmetry** if every point in the configuration has a corresponding point such that the centre of symmetry bisects the line between the points.

characteristic The **integer** part of a **logarithm**. For example:

$$0.056 = 5.6 \times 10^{-2},$$
$$\log_{10}(0.056) = \log_{10}(5.6 \times 10^{-2})$$
$$= \log_{10}(5.6) + \log_{10}(10^{-2})$$
$$= -2\log_{10}(10) + \log_{10}(5.6)$$
$$= -2 + 0.7482$$
$$= -\bar{2}.7482.$$

The $\log_{10}(0.056)$ has a numerical value of -1.2518, but is written $-\bar{2}.7482$. The $\bar{2}$ is the characteristic of the logarithm. See also **mantissa**.

chord A line joining two points on a curve.

chord: *examples of chords*

circle The **curve** on a **plane** formed by the set or **locus** of all points that are **equidistant** from a fixed point. The fixed point is the centre of the circle and the distance of the perimeter or **circumference** from the centre of symmetry is the **radius**.

The **irrational number** π (**pi**) is defined as the ratio of the length of the circumference of a circle to its **diameter**. Therefore the circumference C and radius R of a circle are related by the following expression:

$$\pi = C/2R, \text{ so}$$
$$C = 2\pi R.$$

Circles have line and rotational **symmetry** of infinite **order** about the centre of symmetry.
The circle is the plane figure that encloses the maximum **area** for a given perimeter.

circumference The perimeter of a circle.

circumscribe To surround a geometrical figure with another so that the two are in contact but do not intersect.
　　A **circle** may be circumscribed to a **polygon** if all the **vertices** of the polygon lie on the circle. For example, the circumscribed circle to a **triangle** is shown in the diagram on the next page. A polygon that is circumscribed by a circle is a **cyclic polygon** (see **cyclic quadrilateral**). See also **incircle**.

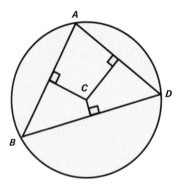

circumscribe: *the circumcircle to a triangle. The perpendicular bisectors of the sides meet at the centre of the circle*

class interval An **interval** of **data values**. For example, the heights h (in cm) of children in a class may at first be represented as follows:

125 126 124 165 164 150 155 130 123 139 140 150
150 166 123 134 156 167 155 137 122 167 164 164

These data values may be more conveniently represented by grouping them into class intervals of 120 cm $< h \le$ 125 cm, 125 cm $< h \le$ 130 cm, etc. and representing them in a **grouped frequency table**. The frequency table for this data and the corresponding **bar graph** are given in the table and diagram.

height (h) cm	number of children
$120 < h \le 125$	5
$125 < h \le 130$	2
$130 < h \le 135$	1
$135 < h \le 140$	3
$140 < h \le 145$	0
$145 < h \le 150$	3
$150 < h \le 155$	2
$155 < h \le 160$	1
$160 < h \le 165$	4
$165 < h \le 170$	3

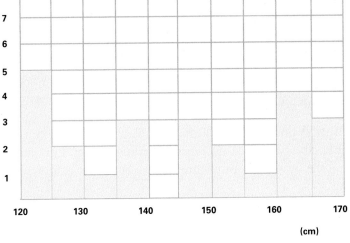

class interval: *a grouped frequency table for the above data*

closed   **1.** A **set** is said to be closed under a mathematical **operation** if the combination (under the operation) of any two members of the set always results in another member of the set. For example the set N of all **integers** is closed under the **addition operation**; that is, any two integers when combined under addition always yield another integer.

2. Describing a **curve** that has no end points.

3. A closed **interval** on the **real number** line is a set of all numbers x that lie in the interval defined by **inequalities** of the form:

$$a \leq x \leq b,$$

where a and b are real numbers. For example, the closed interval $\{x: 1 \leq x \leq 3\}$ is the set of all real numbers between, and including, the 'end points' 1 and 3. An interval such as this is denoted by [1,3].

cluster sampling See **sampling**.

coefficient **1.** A numerical factor prefixed to an unknown quantity in an algebraic term. For example, in the expression:

$$4x^2 + 7x^3y^2 + 14y,$$

the coefficient of x^2 is 4, the coefficient of x^3y^2 is 7, and the coefficient of y is 14.

2. In science and engineering, a number that serves as a measure of some property or characteristic. For example, the *coefficient of restitution* in applied mathematics describes how elastic a collision between two bodies is. The coefficient has a value of 1 for a perfectly elastic collision. A coefficient of restitution of 0 corresponds to a coalescing of the two bodies.

collinear Describing a **set** of **points** that all lie on the same straight **line**.

column matrix See **matrix**.

combination A possible set of a given number of things selected from a given number, irrespective of arrangement within the set. For example, the full list of all the combinations of three letters of the set $\{a, b, c, d\}$ is: *abc, abd, acd, bcd*.

One may think of the combining of these letters as placing the letters into two boxes; one (the one we are interested in) containing three letters, the other containing only one letter. The possible ways of arranging three letters chosen from the set into the first box is given by: 3 ways (first choice) × 2 ways (second choice) × 1 way (last choice), which gives 6 ways; 3 × 2 × 1 is often referred to as **factorial** 3, which is written 3!. The number of ways of arranging the one remaining letter into the last box is only 1, which can be written as 1!. The total number of arrangements for the two boxes combined is therefore 3! × 1!.

Let the total number of ways of choosing sets of one and three letters from a set of four be M. The total number of arrangements of one and three letters from a set of four is therefore just $M \times 3! \times 1!$. This must be the same as the total number of ways that one can arrange all four letters since any combination that we may choose must figure in this set of possible combinations. The possible ways of arranging all four is given by: 4 ways (first choice) $\times$ 3 ways (second choice) $\times$ 2 ways (third choice) $\times$ 1 way (final choice), which is 24 ways or 4!. Therefore, $M \times 3! \times 1! = 4!$, and thus:

$$M = \frac{4!}{3! \times 1!}$$

More generally, the number of combinations of r objects that can be made from a set of n objects is given by:

$$C_r^n = \frac{n!}{r! \times (n-r)!}$$

The values of C_r^n for different values of r are given by the entries in the appropriate **row** of **Pascal's triangle**. See **permutation**.

combination of transformations The successive application of the same or different **transformations** onto a **column matrix**. See **matrix transformations**.

combined probability The **probability** of an **event** that is the combination of two or more constituent events. For example, the probability that the outcome of tossing a coin twice results in two heads is a combination of the probability of obtaining a head for the first toss and a head for the second toss, that is $1/2 \times 1/2 = 1/4$. This may be illustrated by a simple **tree diagram** as shown:

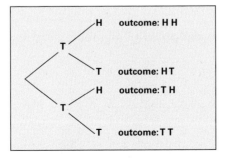

combined probability: *example of a tree* **diagram**

common denominator An **integer** that is a shared multiple of the **denominators** of two or more **fractions**.

In preparation for the **addition** or **subtraction** of fractions with unlike denominators, the fractions are usually modified into equivalents containing a common denominator. For example:

$$\frac{1}{2} + \frac{1}{3} = \frac{3}{6} + \frac{2}{6} = \frac{5}{6}$$

In this case, 6 is the common denominator. See also **lowest**.

common difference The **difference** between successive terms in an **arithmetic sequence**. See also **sequence**.

common formula The general **solution** of a **quadratic equation** of the form:

$$ax^2 + bx + c = 0$$

is given by the common formula:

$$x = \frac{-b \pm \sqrt{b^2 - 4ac}}{2a}$$

common logarithm A **logarithm** to the **base** 10.

common ratio The **ratio** of two successive terms in a **geometric sequence**. See also **sequence**.

commutative A mathematical **operation** is said to be commutative for all the members of a given set {$a, b, c, \dots\dots$} if it has the property $ab = ba$ for all members of that set. For example, **multiplication** on the **real numbers** is commutative; that is, $ab = ba$, where $a, b \in R$. By contrast, **division** on the real numbers is not commutative; that is, $a \div b \neq b \div a$, where $a, b \in R$.

complement The complement C' of a **set** C is the set of all **elements** that do not belong to C:

$$C' = \{x : x \notin C\}.$$

For example, in the domain of all **integers**, if C is the set of all **even** numbers then its complement C' is the set of all **odd** numbers.

complementary angles A pair of **angles** whose **sum** is 90°. See illustration opposite.

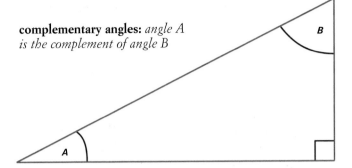

complementary angles: *angle A is the complement of angle B*

complementary events Two **events** are said to be complementary if the occurrence of one event excludes the occurrence of the other. For example, when tossing a coin the occurrence of a 'Head' is said to be the complementary event to obtaining a 'Tail'.

completing the square A method of solving a **quadratic equation** by reducing it to the form:

$$(x + a)^2 = b.$$

For example, consider the following solution for the equation $x^2 - 10x + 1 = 0$:

$$x^2 - 10x + 1 = 0,$$
$$x^2 - 10x = -1,$$
$$x^2 - 10x + 25 = -1 + 25,$$
$$(x - 5)^2 = 24 \Rightarrow x - 5 = \pm \sqrt{24} = \pm 4.9.$$

Either $x - 5 = 4.9$ or $x - 5 = -4.9$; that is, either $x = 9.9$ or $x = 0.1$. See also **Appendix 2**.

complex numbers Numbers of the form $a + ib$, where $i = \sqrt{-1}$ and $a, b \in \mathrm{R}$. The complex numbers are an extension of the **real numbers** originally introduced to deal with equations of the form $x^2 + 4 = 0$, which has no real number solutions:

$$x^2 + 4 = 0,$$
$$x^2 = -4 \Rightarrow x \pm\sqrt{-4} = \pm\sqrt{-1} \times \sqrt{4},$$
$$x = \pm i \times 2 = \pm 2i.$$

When using the **common formula** to solve a general **quadratic equation**, the **discriminant** $d = (b^2 - 4ac)$ may sometimes take on negative values. With complex numbers we are able to say that every quadratic equation has a solution, but that not all these solutions are real numbers.

component The **projection** of a **vector** onto a particular direction (see illustration below). For example, a vector in 3 **dimensions** may be projected onto three mutually **perpendicular** directions. The three vectors resulting from these projections are called components of the original vector. They may be combined under vector addition (see also **addition**) to form the original vector.

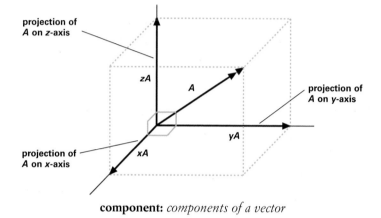

component: *components of a vector*

composite function A function that may be considered to be the combination of the effects of two or more other functions. For example, the function $h(x) = (x^3 + 5)^2$ may be thought of as the combination of two other functions f and g such that:

$$h(x) = f{\circ}g(x),$$
$$h(x) = f{\circ}g(x) = f(g(x))$$
$$g{:}x \rightarrow (x^3 + 5)$$
$$f{:}y \rightarrow y^2$$
$$h(x) = f{\circ}g(x){:}x \rightarrow (x^3 + 5) \rightarrow (x^3 + 5)^2.$$

$h(x)$ may be considered to be the combined action of f then g on the variable x. Note that the **order** of combination of the functions in this way is in general not **commutative**; that is:

$$f{\circ}(g(x)) = (x^3 + 5)^2 = h(x).$$

Now $g{\circ}f(x) = g(f(x)) = x^6 + 5$, since $f(x) = x^2$:

$$g(f(x)) = (x^2)^3 + 5 = x^6 + 5,$$

so $f{\circ}g(x) \neq g{\circ}f(x)$, $\forall x$.

compound angle formulae Formulae that give the values of **trigonometric functions** of the **sum** or **difference** of two **angles**. For example:

$$\sin(30° + 40°) \neq \sin(30°) + \sin(40°);$$

in fact $\sin(30° + 40°)$ and other functions of compound angles may be derived from the formulae shown:

$$\sin(A + B) = \sin(A)\cos(B) + \cos(A)\sin(B)$$
$$\sin(A - B) = \sin(A)\cos(B) - \cos(A)\sin(B)$$
$$\cos(A + B) = \cos(A)\cos(B) - \sin(A)\sin(B)$$
$$\cos(A - B) = \cos(A)\cos(B) + \sin(A)\sin(B)$$
$$\tan(A + B) = [\tan(A) + \tan(B)]/[1 - \tan(A)\tan(B)]$$
$$\tan(A - B) = [\tan(A) - \tan(B)]/[1 + \tan(A)\tan(B)]$$

See also **Appendix 2**.

compound interest See **interest**.

compound measure A **unit** consisting of some combination of **base units**. For example the unit for **speed** is a combination of the base units for distance (metres) and time (seconds):

$$\text{speed} = \text{distance/time, unit (metres/second).}$$

concave Describing a **curve** or surface that may be described as hollow with respect to a given point of reference. For example, the **parabolic function** shown in the diagram overleaf may be described as concave with respect to the positive z-axis.

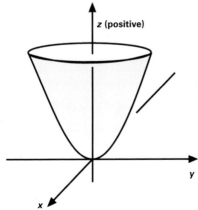

concave: *the surface shown is concave with respect to the z-axis*

concentric circles **Circles** with different radii whose **centres** coincide.

concurrent Possessing a common **point**. A **set** of lines are said to be concurrent if they all meet in a single point. For example, the set of all **diameters** of a **circle** are concurrent because they all share a common point, which is the **centre** of the circle.

conditional probability If the **probability** of an **event** happening depends on whether or not another event took place, then the probability is said to be conditional.

For example, consider a bag of sweets containing 5 red sweets, 5 green sweets, and 5 yellow sweets. A child pulls single sweets out at random and eats them. The probability that the owner will pull out and eat two red sweets in the first two **trials** is a **combined probability** of the two events. However, the probability of the second event is a conditional probability of the first, because the child does not replace the sweets.

The method of calculating this conditional probability is illustrated in the **tree diagram**.

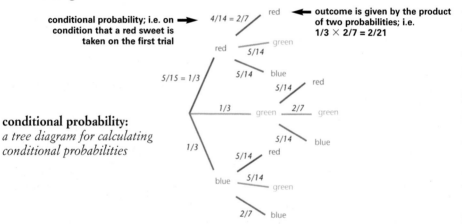

conditional probability; i.e. on → condition that a red sweet is taken on the first trial

outcome is given by the product of two probabilities; i.e. 1/3 × 2/7 = 2/21

conditional probability: *a tree diagram for calculating conditional probabilities*

cone A solid figure with a **vertex** and a **base** that is either a **circle** or an **ellipse**. A *right cone* is one in which the **axis** of **symmetry** is **perpendicular** to the base.

For a cone with a circular base, the volume V of the cone is given by the formula:

$$V = \pi r^2 h/3,$$

where r is the base radius and h is the height.

The curved surface A of a right cone is given by:

$$A = \pi r l,$$

where l is the slant length.

congruent Describing plane figures that have the same shape and size but not necessarily the same orientation.

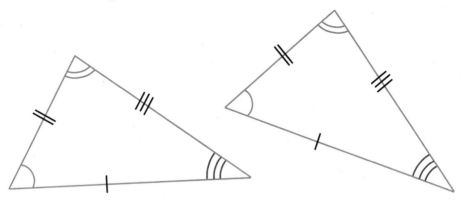

congruent: *the two triangles shown are congruent*

conjugate angle The conjugate angle of an angle $\theta°$ is $(360° - \theta°)$; that is, two angles are conjugate if their **sum** is $360°$.

constant A quantity that remains fixed in value. For example, in the mathematical expression $y = 15x + 7$, 15 and 7 are constants whilst x and y are **variables**. See also **Appendix 2**.

construction Additional points or lines that aid the construction of a mathematical figure.

continuous Describing **data** that can take any value within a given range. For example, the heights of pupils in a class is continuous data. In contrast, the number of pupils in different classes is **discrete** data.

converge A **sequence** is said to converge if the terms get closer and closer to a particular number or **limit**. For example, the sequence:

$$3\tfrac{1}{2}\ \ 3\tfrac{1}{4}\ \ 3\tfrac{1}{5}\ \ 3\tfrac{1}{6}\ \ ...$$

converges since the terms get closer and closer to 3. We say that this sequence converges to 3. See also **diverge** and **oscillate**.

convex Curving outwards. A curve or surface that bulges towards a given **point** of reference is said to be convex.
   Plane figures may also be described as convex. If one can join any two points within the boundaries of the figure by a straight line that itself remains entirely within the figure, then the figure is convex (see illustration overleaf).

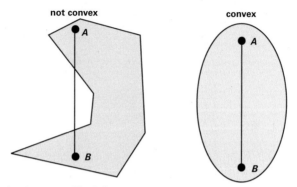

convex: *the figure on the left is not convex because the line AB does not lie entirely within the figure. The figure on the right is convex.*

coordinate One of a **set** of numbers that fix the positions of points in a **plane** or in space. **Cartesian coordinate** systems are most commonly used.

coplanar Lying in the same **plane**. Sets of **points** or lines lying in the same plane are said to be coplanar.

correlation The relationship between the **variables** on a **scatter graph**. If the **data points** exhibit any correlation, a **line of best fit** may be drawn through them. For **positive** correlation the data points must be clustered around a line with a positive **gradient**. A scatter graph of the height of pupils in a class against their weight might exhibit a positive correlation. For **negative** correlation the data points must be clustered around a line with a negative gradient. Data may also exhibit no correlation; that is, there is neither positive nor negative correlation.

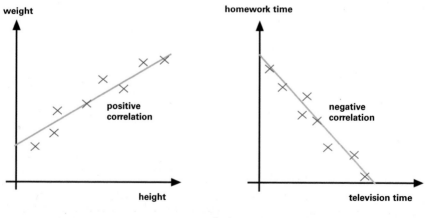

correlation

corresponding points, sides, and angles The full complement of corresponding properties for two or more **similar** figures. For example, the two **triangles** in the diagram have corresponding points, sides, and angles.

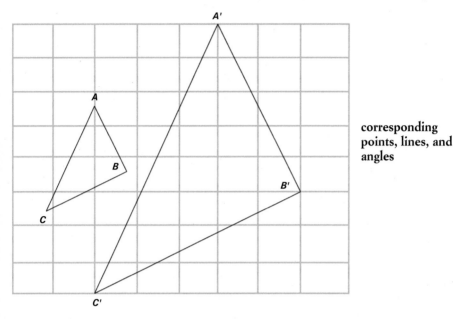

corresponding
points, lines, and
angles

cosine A trigonometric function. In a **right-angled** triangle the cosine of an **angle** is calculated by the **ratio**:

$$\cos (\theta) = \text{adjacent/hypotenuse}$$

where the **adjacent** and the **hypotenuse** are shown in the diagram. See also **Appendix 1**, and **2**.

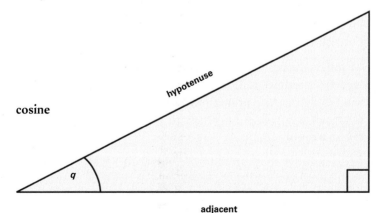

cosine rule For the **triangle** ABC the cosine rule is given by the following expressions:

$$a^2 = b^2 + c^2 - 2bc\cos(A)$$
$$b^2 = a^2 + c^2 - 2ac\cos(B)$$
$$c^2 = a^2 + b^2 - 2ab\cos(C)$$

cosine rule

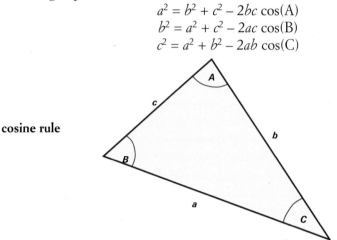

count To assign objects in a **set** in a one-to-one fashion with the names of the positive **integers**. The **positive** integers of **natural** numbers are often called *counting numbers*.

critical path The minimum time and resources needed to complete a set of tasks (which may or may not depend on each other) may be determined by constructing a planning **network**. The path through the planning network that gives this minimum in time and resources is called the critical path.

cross-multiplying A process of manipulation of an algebraic expression. See **Appendix 2**.

cross-section The **plane** figure resulting from taking a section through a **solid**, often in a plane perpendicular to the **axis** of symmetry of the solid. (see diagram opposite)

cube 1. A **polyhedron** with six square **faces**. The cube is a member of a family of solids sometimes referred to as **Platonic solids**.
2. To raise a number by the **power** of three. For example, 'two cubed' is written: 2^3, which is $2 \times 2 \times 2 = 8$.
See also **cubic, cuboid**.

cube root A number that raised to the power of 3 yields a given number. The **inverse operation** to raising a number by the **power** of three (cubing the number) is to take the cube root of the result. The cube root has the symbol $3\sqrt{}$. For example, $3\sqrt{8} = 2$.

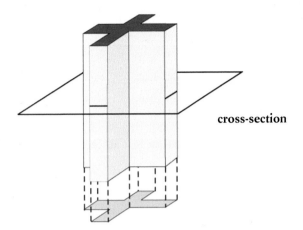

cross-section

cubic 1. A cubic function is a **polynomial** of the third **degree**. For example, the polynomial:

$$ax^3 + bx^2 + cx + d$$

is a polynomial in x where a, b, c, and d are **constants** of variation.
2. A cubic metre, cubic centimetre, etc., are units of **volume**.

cuboid A **solid** with six **rectangular** faces, the opposite **faces** being equal in **dimension**.

cumulative frequency In **statistics**, a method of grouping the **frequency** of values of some **variable** by **summing** the frequencies below certain values of the variable. A cumulative frequency graph is called an **ogive**. Data from the entry for **class interval** may be used to illustrate the concept of cumulative frequencies.

The ogive for this data is given in the diagrams below and overleaf.

Class interval	Frequency	Cumulative frequency
$120 < h \le 12$	5	5
$125 < h \le 13$	0	7
$130 < h \le 13$	5	8
$135 < h \le 14$	0	11
$140 < h \le 145$	0	11
$145 < h \le 150$	3	14
$150 < h \le 155$	2	16
$155 < h \le 160$	1	17
$160 < h \le 165$	4	21
$165 < h \le 170$	3	24

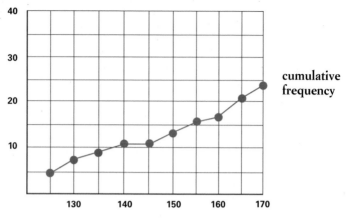

cumulative frequency

curve A line, either straight or continuously bending.

cyclic quadrilateral A four-sided plane figure whose **vertices** lie on the **perimeter** of a **circle**. Two relationships between the angles of a cyclic quadrilateral follow.
(1) The **sum** of opposite **angles** of a cyclic quadrilateral always equals 180°.
(2) The exterior angle of a cyclic quadrilateral equals the opposite interior angle.

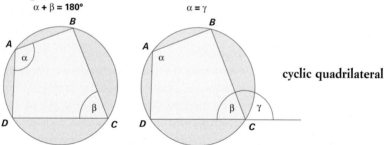

cyclic quadrilateral

cylinder A **solid** that has one **axis** of **symmetry** along which it has a uniform circular **cross-section**.

The **volume** V of a cylinder of radius r and height h is calculated by:

$$V = \pi r^2 h,$$

that is, the area of the cylinder's cross-section (πr^2) multiplied by its height (h).

The curved surface area A of the same cylinder is calculated by:

$$A = 2\pi rh;$$

that is, the **circumference** of the circular cross-section ($2\pi r$) multiplied by the height (h).

D

data Information, especially numerical information, used in **statistics**. **Continuous** data can take any value within a given **range**. **Discrete** data can take only particular values, usually **integer** numbers. For *grouped discrete data*, the values of the **cumulative frequency** are the number of data values that are less than or equal to the upper boundaries of each **class interval**. Data such as this is tabulated on **grouped frequency tables**.

decagon A **polygon** with ten sides.

decay rate A successive reduction in the amount of a particular quantity in **unit** time.

A decreasing quantity is said to have a *constant decay rate* if its decay occurs by a **subtraction** at regular intervals of a fixed amount from a starting value.

A decreasing quantity is said to have an *exponential decay rate* if its decay occurs by a **multiplication** at regular intervals of a fixed factor (of **magnitude** less than 1) with a starting value.

A constant decay rate means that no matter when one observes the decay, the rate of decay will always be the same. An exponential decay rate is variable and depends on the time that the observation of the decay is made. Variable decay rates often take the form of negative **exponential functions** of the base e.

decimal **1.** A system of writing numbers as ten and **powers** of ten. See also **base**.
2. A **fraction** expressed by continuing the ordinary number system to the base 10 into **negative** powers of 10 (a point being used to mark off the fraction from the whole number). For example, the fraction:

$$\frac{567}{1000}$$

may be expressed as the following decimal **sums**:

$$\frac{567}{1000} = 5 \times \frac{1}{10} + 6 \times \frac{1}{100} + 7 \times \frac{1}{1000}$$

degree 1. A unit of angular measurement (see **angle**). One complete **rotation** constitutes 360 degrees, which is written 360°.
2. The degree of a **polynomial** expression in one **variable** is the highest **power** to which the variable is raised, e.g. $x^2 + 2x +1$, is an expression of degree 2 in x.
3. For **products** containing more than one variable, the degree is the **sum** of the powers of the variables in the product, e.g. $x^2y^4z^5$, is said to be a product of degree 11, but of degree 4 in y.

denominator The number under the line in a **fraction**. See also **common denominator** and **lowest**.

density The ratio of the **mass** of an object to its **volume**; that is:

$$\text{density} = \text{mass} \div \text{volume}.$$

dependent variable See **variable**.

determinant A square array of numbers associated with a **matrix**. The determinant of a 2 × 2 **matrix** is defined as the **difference** between the product of the numbers on the leading diagonal and the **product** of the numbers on the other **diagonal**. For example, the **elements** of the leading diagonal in the the following matrix are underlined:

$$\begin{pmatrix} \underline{12} & 15 \\ 3 & \underline{7} \end{pmatrix}$$

The difference of the product of these two diagonals is therefore: $84 - 45 = 39$. The determinant of this matrix is therefore 39.
    The determinant can be defined only for square matrices.

deviation 1. The distance of a **data** value from the **mean** value is called its *deviation from the mean*.
2. It is often important to have a quantity that measures the **average** deviation or *mean deviation* from the mean data value. However, having chosen the mean value as a point of reference, the subsequent deviations will have a **directed** nature. To make the values more useful one just considers the **absolute value** or **modulus** of the **difference** of each data value from the mean and ignores the **sign**. The mean size of deviations can then be calculated as follows:

$$\sum_{x} \frac{|x - \bar{x}|}{n}$$

where x are the data values, $\bar{x}$ is the mean value of x, and n is the number of data values.

3. The *standard deviation*, often denoted by the Greek letter σ. It is given by the formula:

$$\sqrt{\frac{\sum_{x}(x - \bar{x})^2}{n}}$$

See also **variance**.

diagonal **1.** A line joining any two nonadjacent vertices of a **polygon**. The relationship between the number of vertices in a polygon and the corresponding number of diagonals is discussed as an example in **Appendix 3** (the example discusses connections between **nodes** which may be thought of as vertices).

2. The *leading diagonal* in a **matrix** is the diagonal line of numbers from the element in the top left of the matrix to the element in the bottom right of the matrix; that is, in the following matrix:

$$\begin{pmatrix} \underline{a} & b \\ c & \underline{d} \end{pmatrix}$$

underlined elements constitute the leading diagonal. The elements b and c constitute the other diagonal.

diameter A **chord** of a **circle** that passes through the **centre** of the circle. The length of the diameter of a circle is twice the length of the **radius** of the circle.

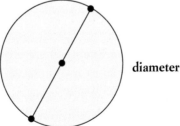

diameter

difference **1.** The result of the **subtraction** of two numbers; i.e., $a - b$ where a, b ∈ R.

2. The *difference of two squares* is the common name given to the following factorization:

$$a^2 - b^2 = (a - b)(a + b).$$

difference method A method of deducing the next term in a sequence of numbers. For example, the **sequence** 12, 14, 22, 36, 56, ... can be written as follows:

12		14		22		36		56		82	
	2		8		14		20		26		(1)
		6		6		6		6			(2)

In line (1) the difference between successive terms is listed. A pattern is now visible in line (2) which is exploited to show the difference of 6 between successive differences.

digits The set of **numerals** that are used in the representation of a number. For example, in the **decimal** system (**base** 10) the numerals 0, 1, 2, 3, ... 9, in combination with a possible decimal point, are used to represent the digits of any number.

digital Describing measuring devices that represent measurements by digital code (usually in terms of ones and **zeros**). See also **analogue**.

dimension 1. A particular direction or extension into space. A **volume** of space is therefore three-dimensional; that is, one needs to specify a length, a breadth, and a height to calculate the volume.
2. All physical quantities can be represented as a combination of a set of arbitrarily chosen base quantities called dimensions: length, mass, time, amount of substance, luminosity, and electric current. For example, **volume** has the dimensions length × length × length (L^3), **speed** has the dimensions of length ÷ time (LT^{-1}), and **acceleration** has the dimensions length ÷ (time × time) (LT^{-2}).

directed Describing numbers that are assigned as **positive** or **negative**, where the positive numbers are represented along a line relative to an origin which represents the value 0, and negative numbers are similarly represented along the reverse direction.

direct proportionality See **proportionality**.

discrete Describing quantities that can be measured by a counting process. Discrete **variables** vary in a discontinuous fashion. See also **continuous**.

discriminant When using the **common formula** to solve a **quadratic** equation of the form:

$$ax^2 + bx + c = 0,$$

the quantity $b^2 - 4ac$ is called the discriminant. The type of solutions for the quadratic may easily be determined by considering the **sign** of the discriminant. For example, if:

$b^2 - 4ac > 0$, the quadratic has real **roots**

$b^2 - 4ac = 0$, the roots of the quadratic coincide,

$b^2 - 4ac < 0$, the roots are **complex numbers**.

disjoint Describing two or more **sets** that do not intersect; that is, the **intersection** of these sets is 0, the **empty set**. For example, the set of all **odd** numbers and the set of all **even** numbers are disjoint; i.e. the two sets have no **members** in common.

dispersion The way in which **data** are distributed. There are many ways of measuring the dispersion of data. The following quantities which are derived from the data are some of the possible ways of considering dispersion: **deviation**, **mean** deviation, **variance**, and **standard deviation**.

displacement A **vector** quantity representing a change in position. See also **magnitude**.

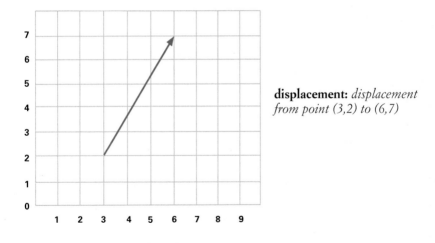

displacement: *displacement from point (3,2) to (6,7)*

distance–time graph A graph on which distance is measured on the **vertical axis** and time is measured on the **horizontal** axis. The slope or **gradient** at any given time is the instantaneous **speed** of the particle whose motion is represented by the graph. The steeper the slope, the greater the speed.

distribution The manner of allotment of **data** values associated with the **frequencies** of a statistical **variable**. In **statistics** there are many different types of distribution. The *normal distribution* results from an unbiased **sample**

of **continuous** data such as measurements of height and weight. It is **symmetrical** about the **mean** which means that the mean = **mode** = **median**. See also **skew**.

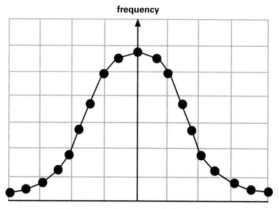

distribution: *the normal distribution*

distributive Consider two **binary** operations ∗ and • defined on a **set** S. The operation ∗ is said to be distributive over • if it satisfies expressions of the form:

$$x * (y • z) = (x * y) • (x * z),$$

where x, y, z ∈ S. For example in the **arithmetic** of **real numbers,** the **multiplication** operation is said to be distributive over **addition**:

$$4 \times (5 + 3) = (4 \times 5) + (4 \times 3).$$

Note that the converse is not true; that is,

$$4 + (5 \times 3) \neq (4 + 5) \times (4 + 3).$$

diverge A **sequence** in which the terms get progressively larger and larger is said to diverge. For example, the sequence 1, 10, 100, 1000, 10 000 … diverges since the terms tend to an **infinite** size. See also **converge** and **oscillate**.

dividend A number that is to be divided by another. For example, in the expression 7654 ÷ 3, the number 7654 is the dividend. See also **division** and **divisor**.

division One of the basic operations of **arithmetic**. Division is the **inverse** operation to **multiplication**; that is

$$A \times B \div B = A.$$

The division of one number by another may also be represented by using the **quotient** symbol. For example, $A \div B$ may be written as A/B. See also **dividend**, **divisor**, and **Appendix 2**.

divisor   The number that is to divide the **dividend** in a division. For example, in the expression $7654 \div 3$, the number 3 is the divisor.

dodecagon   A **polygon** having twelve sides and twelve interior angles.

dodecahedron   A **polyhedron** having twelve faces. A **regular** dodecahedron has twelve regular **pentagons** as faces and is a member of a family of solids sometimes referred to as the **Platonic solids**.

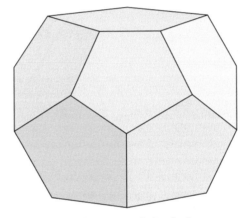

dodecahedron: *a dodecahedron*

domain   The **set** whose elements act as inputs to a **function**. For example, in the circle of radius R:

$$x^2 + y^2 = R^2,$$

the domains of the variables x and y are:

$$-R \leq x \leq R,$$
$$-R \leq y \leq R.$$

To see this more clearly, rearrange the equation of the circle to the form $y = f(x)$; that is,

$$y = \pm\sqrt{(R^2 - x^2)}.$$

In this form we can think of producing two y values every time we choose an x value. However, we cannot just choose any x value since any x value with a **modulus** greater than R would lead to **roots** made up of only

imaginary parts for $y(x)$. Therefore, the modulus of x must be less than or equal to R; that is, the domain of x for the function $y(x)$ is:

$$-R \leq x \leq R.$$

dual Describing two geometrical figures that can be obtained from each other by the interchange of positions of various parts of the figures. For example, in the **set** of five **Platonic solids** each **polyhedron** is the dual of another in the set. To see this one need only consider the result of interchanging the vertices with faces or vice versa for any of these solids.

In the case of the **tetrahedron**, an interchange of faces and vertices leads to another tetrahedron; the tetrahedron is therefore said to be *self-dual*.

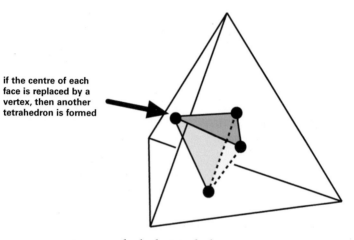

if the centre of each
face is replaced by a
vertex, then another
tetrahedron is formed

dual: *the tetrahedron*

In this way the **cube** is easily shown to be the dual of the **octahedron** and the **dodecahedron** is similarly the dual of the **icosahedron**.

E

e   The **irrational** number that is the **base** for natural **logarithms**. It has a value of 2.718 281 828..., which is the limit of the infinite **sum**:

$$1 + 1/1! + 1/2! + 1/3! + 1/4! + ... + 1/n!$$

where $n!$ designates the **factorial** of n.

The number e appears frequently in problems pertaining to **growth** and **decay** rates. In fact e will definitely be involved in any problem where the amount of decay or growth of a quantity over a given time is always the same proportion of the amount that is left. For example, if a child were to eat half of what is left of a chocolate bar each second then the consumption of the bar could be represented by a graph, as shown below.

The subsequent reduction or decay of this bar of chocolate could be approximately written as the decay:

$$A(t) = e^{-0.693t}$$

where $A(t)$ is the amount of the chocolate bar left after t seconds, $t = 1, 2, 3,.......10$, and e is the base of natural logarithms.

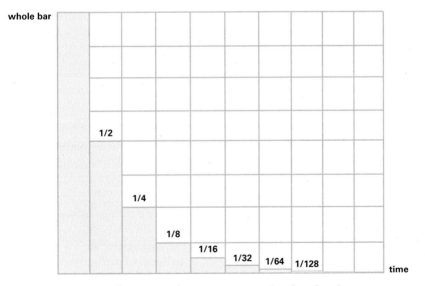

e: *decay curve for consumption of a chocolate bar*

edge   The line formed by the **intersection** of two **planes**. For example, the line formed at the join of two **faces** of a **polyhedron** is an edge.

element A **member** of a **set**. The symbol $\in$ is used to express this concept; for example:

$$7 \in \{\text{all odd numbers}\},$$

reads '7 is an element of the set of all odd numbers'.

elevation 1. The *angle of elevation* of an object with respect to an observer is the **angle** between the **horizontal plane** containing the observer and the straight line between the object and the observer.
2. In the two dimensional representation of a solid object, the *front elevation* is the view of the object from the front. Similarly, the *side elevation* is the view of the object from the side.

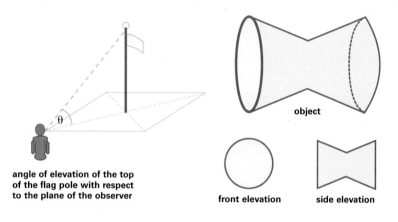

angle of elevation of the top
of the flag pole with respect
to the plane of the observer

object

front elevation   side elevation

elevation

elimination method A method of solving **simultaneous equations** that depends on eliminating one of the unknown quantities. For example, the simultaneous equations:

$$2x + 3y = 13$$
$$x + y = 5$$

may be modified multiplying the second equation by 2 to give:

$$2x + 3y = 13$$
$$2x + 2y = 10$$

Taking the terms in the second equation from the first eliminates the unknown x. This leads to the simple equation $y = 3$, which in turn leads to $x = 2$.

ellipse The **curve** on a **plane** formed by the **locus** of an object moving so that the **sum** of its distances from two fixed points, called *foci*, is **constant**. Such a locus may be constructed using a pair of compasses. If the *foci* chosen are labelled (1) and (2), then the calculations may be tabulated:

Distance from (1):	4.5	4	3	2	1	0.5
Distance from (2):	0.5	1	2	3	4	4.5
Sum of distances:	5	5	5	5	5	5

The compasses may be opened out to these distances and the joining of the crossings of arcs results in the figure shown.

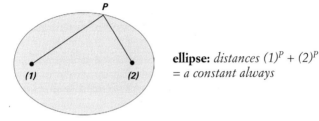

ellipse: *distances* $(1)^P + (2)^P$ = *a constant always*

empty set A **set** containing no **members**. The empty set is denoted by the symbol Ø. For example, the set of all **integers** that are both **odd** and **even** is empty; that is, there are no **integers** that are both odd and even.

enlargement An abstract **transformation** of a **plane** figure or solid object that increases the size of the figure or object by a *scale factor* but leaves the shape **invariant**. For example, in the diagram the smaller triangle has been transformed by an enlargement into the larger triangle. The larger triangle is the **image** of the smaller **object**. The scale factor in this case is the **ratio** of corresponding lengths on the object to those on the image. See also **centre of enlargement** and **scale factor**.

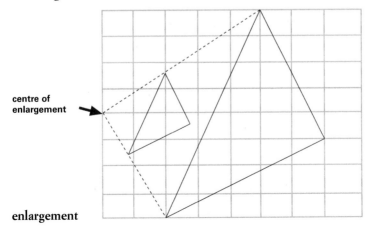

centre of enlargement

enlargement

envelope A line that is **tangential** to each member of a whole family of **curves**. An example of envelopes that are the lines that mark the extremities of **sine** and **cosine** curves. This is illustrated in the diagram shown for the curve $y = \sin(x)$. In fact the lines $y = 1$ and $y = -1$ are envelopes for all sine and cosine curves of the form:

$$y(x) = \sin(ax + b)$$

where a, and b are arbitrary **constants** belonging to the **real numbers**.

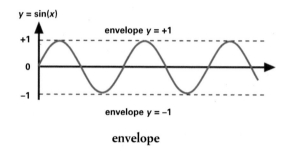

envelope

equal Identical in quantity. The symbol = is used to denote equality. The relationship of equality is an example of an **equivalence relation**.

equally likely events Two **events** are described as equally likely if the **probabilities** of occurrence of the two events are **equal**. For example, when a coin is tossed the outcomes 'Head' and 'Tail' are events that both have a probability of occurrence of 1/2 and are therefore equally likely.

equation A statement of the equality of two quantities or mathematical expressions. An equation may be an expression of identity or an expression that leads to an explicit value for unknown quantities. For example, for x, y, $z \in R$, the equation:

$$x(5y + 7z) = 5xy + 7xz,$$

expresses the identity of the expression $x(5y + 7z)$ with $5xy + 7xz$, which is a consequence of the **distributive** property of **multiplication** over **addition** on the **real numbers**; that is, for any real numbers x, y and z the above expression holds.

The equation: $4x + 5 = 9$, is an equation which leads to an explicit value of 1 for x. This equation is an example of a **linear** equation that is conditional on the value of x. Other examples of common types of simple equations include **quadratic** and **simultaneous equations**. See also **Appendix 2**.

equator An imaginary circle passing round the Earth, **equidistant** from North and South poles. The equator is the **great circle** from which positions of **latitude** are referenced. For example, a latitude of $\theta°$ North ($\theta°$ North of the equator) is shown in the diagram.

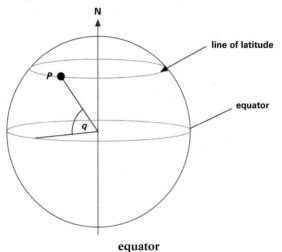

equator

equidistant **Points** that lie the same distance from a reference point are all said to be equidistant from the reference point.

equilateral Having all sides **equal** in length. For example, an equilateral **triangle** is a triangle whose sides are of equal length. This condition means that, for the triangle, all three interior angles must be equal to 60°. This of course is not necessarily the case for all equilateral **polygons**.

equilibrium (*pl.* **equilibria**) A state of balance in which opposing forces or tendencies neutralize each other. A particle at a point of equilibrium may be displaced from that point with consequences which are dependent on the forces and tendencies in the local region. For example, the three diagrams show initial positions of marbles on various surfaces:

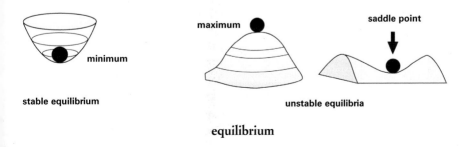

equilibrium

For *stable equilibrium* the marble must be in a local **minimum** with respect to the surface. Any small displacement from the local minimum will result in the marble rolling back to the minimum.

For *unstable equilibrium* the marble can either be on a local **maximum** or a local *saddle point* (so named on account of its shape). Small displacements from either of these points can result in the marble rolling away from these points.

equivalence relation On a **set** S, an equivalence in a particular property P relates two members of S that share the property P. A relationship $*$ is said to be an equivalence relation of P on S if it has the following three properties:
(1) It is **reflexive**; that is, $\forall a,b \in S$, $a * a$; for all a in S, a is equivalent to a with respect to the property P.
(2) It is **symmetric**; that is, $\forall a,b \in S$, if $a * b$ then $b * a$; for all a and b in S, if (with respect to P) a is equivalent to b then b is equivalent to a.
(3) It is **transitive**; that is, $\forall a,b,c \in S$, if $a * b$ and $b * c$ then $a * c$; for all a, b and c in S, if (with respect to P) a is equivalent to b and b is equivalent to c then a is equivalent to c.

For example, the equality relation (**equal**) on the **real numbers** is an example of an equivalence relation because:
$$\forall a \in R \; a = a,$$
$$\forall a,b \in R \; a = b \text{ then } b = a,$$
$$\forall a,b,c \in R \; a = b \text{ and } b = c \text{ then } a = c.$$

equivalent fractions **Fractions** that by a process of **cancellation** can be shown to represent the same **rational number**. For example, 3/4, 6/8, 9/12, and 21/28 are all equivalent fractions.

equivalent ratios Equivalent ratios may be formed by **multiplication** (or **division**) of both sides of the **ratio** by the same number. For example, the ratios 2:3, 16:24, and 32:48 are all equivalent ratios.

error in measurement When using a measuring device to measure a value to the nearest **unit**, one has to decide whether the measurement is nearer one mark or another; that is, the measurement is within half of the smallest increment measurable on the device. For example, a distance given as 289 km to the nearest km could be between a **lower bound** of 288.5 km and an **upper bound** of 289.5 km.

estimate To obtain an inexact but helpful preliminary **solution** to a problem. An estimation may involve one or more **approximations**

appropriate for the level of **accuracy** of the preliminary calculation. For example, a possible estimate for the calculation:

$$(8.81 \times 74.85)/0.46$$

could be obtained by replacing the numbers with numbers that are easier to work with, for example:

$$(8.81 \times 74.85)/0.46 \cong (9 \times 75)/0.5,$$

giving an estimate of 1350.

Euclidean Pertaining to Euclid, a 3rd-century BC Greek geometrician from Alexandria who set out the principles of **geometry**. The term is often used to describe geometry in ordinary two- and three-dimensional spaces.

Euler's formula See **Appendix 3**.

even Describing **integers** that have 2 as a factor. Even numbers are multiples of 2: i.e. 2, 4, 6, 8, ...

event An occurrence, such as the toss of a coin resulting in a 'Head' or 'Tail'. The measure of the possibility of an event is its **probability**.
(1) *Exhaustive events* account for all possible outcomes. If a set of events is exhaustive then it is certain that one of them will occur.
(2) *Dependent events* lead to **conditional probabilities** because the occurrence of one event is conditional on whether or not the other event has occurred.
(3) *Independent events*: events A and B are independent if the occurrence of A has no influence on the occurrence of B.
(4) *Mutually exclusive events* are events that cannot happen at the same time. For example, the toss of a coin can result either in a 'Head' or a 'Tail' but not both.

expansion The expanded form of a mathematical expression. For example, $3(2x + 4)$ is expanded to $6x + 12$; $(x - 2)(2x + 3)$ is expanded to $2x^2 - 4x + 3x - 6$ and simplified to $2x^2 - x - 6$. These expansions have been possible due to the **distributive** property of **multiplication** over **addition** for these expressions. See **Appendix 2**.

exponent The **power** to which a **base** is to be raised to is represented as a numerical symbol called an exponent. For example, the exponent in 6^2 is 2. See also **square**, **cube**, and **index**.

exponential function **Functions** involving **variables** in the **exponents** of a **base** are called exponential functions.

For example, the equation $y(x) = 2^x$ is an exponential function of x where the base is 2. The table following shows the values of this function for some values of x:

x	−2	−1	0	1	2	3
2x	$2^{-2}=0.25$	$2^{-1}=0.5$	$2^0=1$	$2^1=2$	$2^2=4$	$2^3=8$

For large **negative** x the function is small and **positive**. For large positive x the function is large and positive. $y(x)$ is 1 for $x = 0$. See **Appendix 1**.

All exponential functions with bases other than 2 have similar forms to $y(x) = 2^x$. In particular the function $y(x) = e^x$, which is often known as the *exponential function*, uses the **irrational** number e as the base. Such functions are associated with problems involving growth or decay in natural systems. See also **logarithm**.

extrapolate To use a trend in a set of **data** to predict future data values. For example, in physics the extension of a spring that obeys **Hooke's law** is described as **linear** with respect to the load applied. Therefore a graph of data values for the spring may be extended beyond the data obtained by experimentation. Real springs of course do not obey Hooke's law for all loads; the spring may reach its *elastic limit*.

F

face The **plane** regions that bound a **polyhedron**. For example, a **dodecahedron** has twelve faces.

factor A whole number that **divides** exactly into a given number or polynomial (see **factor theorem**). For example:
(1) The factors of 10 are 1, 2, 5, and 10.
(2) A factor of the expression:

$$10x + 6$$

is 2, since it is possible to write this expression as:

$$2(5x + 3).$$

Such a manipulation is called a *factorization*.
See **Appendix 2**. See also **distributive**.
   A number may be written as a product of factors that are **prime numbers** with the aid of a factor tree. For example, the factor tree for the number 90 is shown.
   The *highest common factor* (HCF) of two numbers is the greatest number that is a factor of both of the given numbers. For example, since 30 may be written as $2 \times 3 \times 5$ and 45 may be written as $3 \times 3 \times 5$, the HCF of 30 and 45 is 3×5; that is, 15.

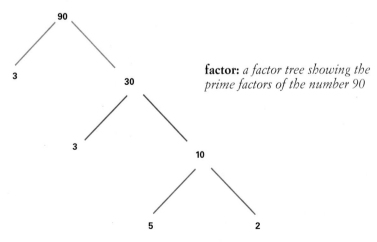

factor: *a factor tree showing the prime factors of the number 90*

factor theorem The theorem that if $P(x)$ is a **polynomial** in x and $P(a) = 0$, then the **factor** $(x - a)$ is a factor of the polynomial; that is:

$$P(x) = (x - a)p(x),$$

where $p(x)$ is another polynomial in x of a **degree** that is one lower than that of $P(x)$. For example, consider the polynomial of degree three:

$$P(x) = x^3 - 6x^2 + 11x - 6.$$

Setting $x = 1, 2,$ or 3 gives:

$$P(1) = 1 - 6 + 11 - 6 = 0,$$
$$P(2) = 8 - 24 + 22 - 6 = 0,$$
$$P(3) = 27 - 54 + 33 - 6 = 0.$$

This means that the expression for $P(x)$ may be simplified by extracting a factor $(x - 1)$; i.e.,

$$P(x) = x^3 - 6x^2 + 11x - 6$$
$$= (x - 1)(x^2 - 5x + 6).$$

A further simplification may be made by extracting a factor $(x - 2)$ from the quadratic $(x^2 - 5x + 6)$; i.e.,

$$(x^2 - 5x + 6) = (x - 2)(x - 3).$$

Thus $P(x)$ may be written as the product of three factors:

$$P(x) = x^3 - 6x^2 + 11x - 6$$
$$= (x - 1)(x - 2)(x - 3).$$

factorial notation A notation that is used to simplify a mathematical **product** of the form $5 \times 4 \times 3 \times 2 \times 1$. Such a product is called $5!$ (pronounced 5 factorial); for example,

$$5! = 5 \times 4 \times 3 \times 2 \times 1 = 120$$
$$4! = 4 \times 3 \times 2 \times 1 = 24$$
$$3! = 3 \times 2 \times 1 = 6$$
$$2! = 2 \times 1 = 2$$
$$1! = 1$$

Note that $0!$ is defined as 1; that is, $0! = 1$.

Fahrenheit scale A temperature scale having the freezing point and boiling point of water at $32°$ and $212°$ respectively.

This leads to a general conversion **formula** to the more common **Celsius scale**:

$$C/5 = (F - 32)/9,$$

where C is the temperature on the Celsius scale and F is the temperature on the Fahrenheit scale.

favourable outcome An outcome that fulfils certain predetermined criteria. For example, the probability of getting a **prime number** when a die is tossed is calculated as follows:

- The possible **equally likely events** are 1, 2, 3, 4, 5, 6. So the total number of possible outcomes is 6.
- The favourable outcomes (die rolls that are primes) are 2, 3, 5. So the number of favourable outcomes is 3.
- The probability of obtaining a prime on a single roll of the die is $3/6 = 1/2 = 0.5$.

Fibonacci sequence A **sequence** of numbers named after the Florentine mathematician Leonardo Fibonacci (c.1170–1250). Each member of the Fibonacci number sequence is formed by the **sum** of the two preceeding members:

$$1, 1, 2, 3, 5, 8, 13, 21, 34, \text{etc.}$$

fixed point If a value is introduced into an **iterative** formula and the same value is the result, then the value is called a fixed point of the iterative process. For example, the fixed point of the iterative formula:

$$u_{n+1} = (u_n/5) + 4,$$

may be found by considering a value x such that when it is introduced into the formula as u_n the result of un+1 is x; that is,

$$x = (x/5) + 4,$$
$$5x = x + 20,$$
$$4x = 20 \Rightarrow x = 5.$$

The fixed point of this iteration is therefore 5.

   Fixed points may occur at points where the iterations like this **converge**. However this is not al-ways the case as the fact that an iterative formula has fixed points does not necessarily mean it converges.

flip movement See **reflection**.

flow diagram A diagrammatic method of illustrating a set of procedures that are to be followed to solve a particular problem. The various procedures are often 'boxed' within different-shaped captions, which signify the nature of the procedure. Three of the most commonly used boxes are illustrated on the next page.

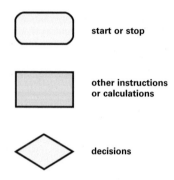

flow diagram: *three typical shapes of boxes commonly used in flow diagrams*

flow-diagram method A method of solving an **equation** of one unknown by following a procedure (often represented as a **flow diagram**) that effectively inverts the algebraic **operations** that make up the equation. For example, the equation:

$$2x - 4 = 1,$$

may be thought of as the flow diagram shown overleaf as *flow 1*.

The solution for x could be obtained by considering the **inverse** of *flow 1* which is shown in *flow 2*.

See also **Appendix 2**.

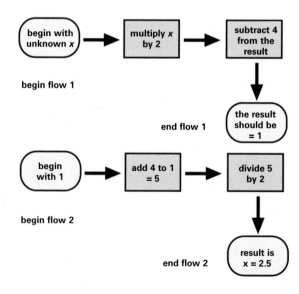

flow-diagram method

foot A **unit** of length. See **unit conversions**.

formula (*pl.* **formulae**) A general algebraic expression for solving problems. For example, the general conversion formula from the **Fahrenheit scale** of temperature to the more common **Celsius scale** is:

$$C/5 = (F - 32)/9$$

where C is the temperature on the Celsius scale and F is the temperature on the Fahrenheit scale.

fraction A part of the whole. For example, the fraction 'two thirds' is denoted by the **quotient** $2 \div 3$ or $\frac{2}{3}$. In this fraction the top number (2) is called the **numerator**; the bottom number (3) is called the **denominator**.
(1) An *improper fraction* is a fraction whose numerator is greater than the denominator. This may lead to an expression of the fraction as a **mixed number**. For example, $7/2 = 3\frac{1}{2}$.
(2) A *proper fraction* is a fraction whose denominator is greater than the numerator. For example, $4/9$.
See also **Appendix 2**.

frequency The commonness of occurrence of an **event** in a given number of **trials**. The *relative frequency* of an event compares the frequency with the number of trials performed. It is the proportion of times the event occurs in a number of trials.
  The relative frequency of an event can be used as an **estimate** of the **probability** of the event. For example, if over a period of one week there were three days that were sunny, the relative frequency of sunny days over this period is just $3/7$. This result may be used as an estimate of the probability of a day being sunny if one assumes that all weeks are identical with respect to the weather. So an estimate of the probability of any particular day being sunny would be $3/7 = 0.4286$ to 4 **significant figures**.

frequency density The height of each bar on a **histogram**. The **area** of each bar represents the **frequency** within the **class interval** represented by the bar. Using:

$$\text{area of bar} = \text{width} \times \text{height},$$

$$\text{frequency} = \text{class interval} \times \text{height of bar}.$$

Hence

$$\text{height of bar} = \text{frequency/class interval},$$

that is,

$$\text{frequency density} = \text{frequency/class interval.}$$

function Complex expressions involving one or more variables may be considered as a relation between two **sets** called the **domain** and the **range**. The relation in a variable x is often represented by the function notation $f(x)$ or 'function of x', which is a convenient way of representing this correspondence between the domain and range. The function f is a relationship between the values x that are elements of the domain of f and the values y that are the set of **images** of x through f. For example, if

$$f(x) = x^2 - 5$$

for $x \in R$, then

$$f(3) = 3^2 - 5 = 4;$$

that is, if 3 is introduced into f then the image of 3 through f is 4. In this way each member of the domain of a function has precisely one image that is a member of the range of the function.

G

gallon A **unit** of **volume**. See **unit conversions**.

geometric mean In a **set** of *n* observations, the nth **root** of the **product** of all the observations. For example, the geometric mean of the numbers 2, 4, and 8 is 4, which is calculated by first finding their product:

$$2 \times 4 \times 8 = 64,$$

then taking the **cube root** of this result to give 4.

geometric sequence See **sequence**.

geometry A branch of mathematics that deals with the properties of **planes**, **points**, straight lines, **curves**, plane shapes, and **solids**.

giga- A prefix with **symbol** 'G' meaning 'one thousand million' (10^9). For example, a mass of five gigatonnes (5 GT) is equivalent to five thousand million tonnes; i.e. 5 000 000 000 tonnes.

glide–reflection transformation A **combination of transformations** composed of a **reflection** and a **translation**. An example of a glide–reflection transformation applied to a square *ABCD* is shown in the diagram. If a **vertex** of the square is represented by a **column matrix**:

$$\begin{pmatrix} x \\ y \end{pmatrix}$$

then the combined transformation to a point may be written down as:

$$\begin{pmatrix} x' \\ y' \end{pmatrix}$$

$$\begin{pmatrix} x' \\ y' \end{pmatrix} = \begin{pmatrix} x+5 \\ -y \end{pmatrix} = \begin{pmatrix} 1 & 0 \\ 0 & -1 \end{pmatrix}\begin{pmatrix} x \\ y \end{pmatrix} + \begin{pmatrix} 5 \\ 0 \end{pmatrix}$$

See **matrix** and **transformations**.

gradient A measure of the steepness of a line or **curve**.

The gradient of a straight line may be obtained by referring the line to a **Cartesian coordinate** system and calculating the **ratio** $\Delta y/\Delta x$, where Δy and Δx are respectively **directed** increments in the y and x directions. Therefore the gradient is positive if the height of the slope increases with increasing x, and negative if the height of the slope decreases with increasing x. Note

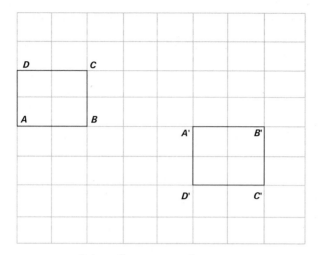

glide–reflection transformation:
a glide–reflection applied to the rectangle ABCD

that, on a straight line, the gradient is the same along the whole length of the line.

The gradient of a continuously bending curve is not necessarily the same at every point. In fact for a continuously bending curve the gradient at a point is also a **function** of the **coordinates** of that point. The gradient at a particular point on a curve may be obtained by calculating the gradient of the **tangent** to the curve at that point. See also calculus in **Appendix 1**.

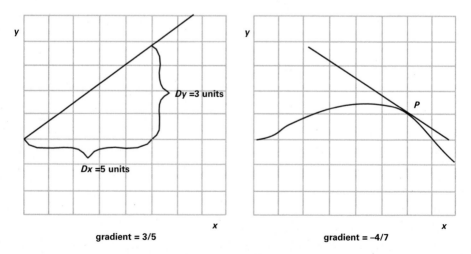

gradient: *the gradient of a line and a curve*

gram A **unit** of mass (see **dimensions**). See **unit conversions**.

graph A diagram consisting of a line or lines representing the relationship between various quantities called **variables.** Graphs are often displayed according to a **coordinate** system (see **Cartesian coordinates**) which quantify the relationship between the variables. Through a coordinate system, there is an intimate relationship between graphs and equations. See also **Appendix 1.**
Statistical information is often displayed in the form of a **bar graph.**

graphical solution of equations A method of solving two simultaneous **equations** by finding out where the graphs of the equations intersect. See **simultaneous equation.**

great circle An imaginary circle drawn on the Earth's surface which has the same **radius** as that of the Earth. For example, the **equator** is a great circle; all lines of **longitude** are parts of great circles (see illustration). See also **small circle.**

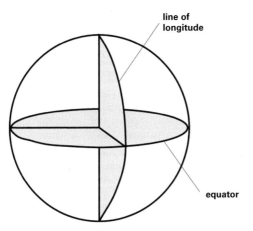

great circle: *examples of great circles on the Earth*

grouped frequency table A method of tabulating **data** by grouping it into **class intervals.** The frequency of data-values that occur within an interval is recorded as the frequency for that class. An example of a grouped frequency table and its corresponding data is given at class interval.

growth rate A successive increase in the amount of a particular quantity in **unit** time.
(1) An increasing quantity is said to have a *constant growth rate* if its growth occurs by an **addition** of a fixed amount to a starting value at regular intervals.

(2) An increasing quantity is said to have an *exponential growth rate* if its growth occurs by a **multiplication** of a fixed factor (of magnitude greater than 1) with a starting value at regular intervals.

A constant growth rate means that no matter when one observes the growth, the rate of growth will always be the same. An exponential growth rate is variable and depends on the time that the observation of the growth is made. Variable growth rates often take the form of positive **exponential functions** of the **base** e.

H

half-life The time taken for the activity of a radioactive isotope to decay to half of its original value; that is, for half of the atoms present to disintegrate. An example of this kind of reduction in a quantity is given in the entry for **e**.

hectare A **unit** of land area. The hectare (ha) is derived from the **are**. Small land areas, such as building plots, are measured in m²:

$$1 \text{ are} = 100 \text{ m}^2,$$

$$1 \text{ ha} = 10\ 000 \text{ m}^2 = 100 \text{ ares}$$

$$100 \text{ ha} = 1 \text{ km2} = 1\ 000\ 000 \text{ m}^2.$$

helix The **curve** formed by the **locus** of a point moving on the surface of a **cylinder** or **cone** making a constant **angle** with the **axis**.

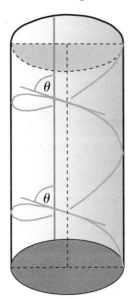

helix: *a helix generated on the surface of a cylinder*

hemisphere Half of a **sphere**.
The **volume** V of a hemisphere of **radius** r is given by:

$$V = {}^2/_3\pi r^3.$$

The curved suface area of a hemisphere is given by:

$$2\pi r^2.$$

The addition of this curved surface area to the circular base leads to a total surface area of:

$$2\pi r^2 + \pi r^2 = 3\pi r^2.$$

heptagon A **polygon** with seven **sides**. The **sum** of all the **interior angles** of a heptagon is 900°. A regular heptagon has sides of equal length.

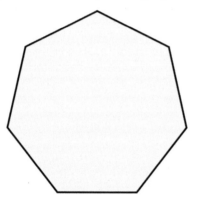

heptagon: *a regular heptagon has all seven sides of equal length*

Heron's formula A **formula** for calculating the **area** of a **triangle** from the lengths of its sides:

$$A = \sqrt{[s(s - a)(s - b)(s - c)]},$$

where A is the area, a, b, and c are the lengths of the sides and s is called the semi-perimeter given by:

$$s = \tfrac{1}{2}(a + b + c).$$

Heron of Alexandria was a 1st century Greek mathematician.

hexagon A **polygon** with six **sides**. The **sum** of all the **interior angles** of a hexagon is 720°. A regular hexagon has sides of equal length.

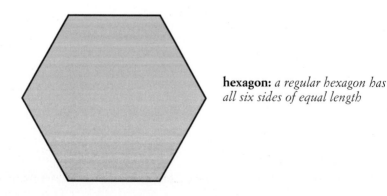

hexagon: *a regular hexagon has all six sides of equal length*

highest common factor See **factor**.

histogram A graph similar to a **bar graph** used for numerical **data**, but in which the bars are joined. The **frequency** of the bar is represented by its **area** rather than its height.

The **vertical axis** of a histogram is **frequency density** or frequency per **class interval** of a given size. The **horizontal axis** acts like a normal scale on a **graph**, with the same distance representing the same number of **units**.

homogeneous polynomial A **polynomial** whose terms are all **products** of the same **degree**.

Hooke's law The principle that the extension of a spring due to a load is in a **direct proportion** to the load applied (provided the spring does not exceed its elastic limit).

horizontal A line is said to be horizontal if it is **parallel** to the Earth's horizon.

hour A **unit** of time. One hour is equal to 60 **minutes**. There are 24 hours in a day.

hypotenuse The longest side of a right-angled **triangle**. The hypotenuse is always opposite the right angle.

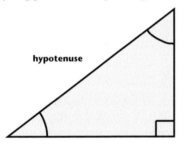

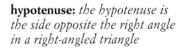

hypotenuse: *the hypotenuse is the side opposite the right angle in a right-angled triangle*

hypothesis A statement of an opinion about an issue. See **Appendix 3**.

For example, the statement 'most cars on the road are red' is an hypothesis. A **survey** could be conducted to test this hypothesis; that is, to find out whether or not the hypothesis is true.

There are usually four steps to making and testing an hypothesis:
Step 1. State the hypothesis based on preliminary observations.
Step 2. Conduct an experiment to test the hypothesis and collect relevant **data**.
Step 3. Analyse the data.
Step 4. Compare the results of the analysis to the original hypothesis.

I

i The **square root** of −1; i.e. $\sqrt{-1} = i$.

identity matrix See **matrix**.

icosahedron A **polyhedron** having twenty **faces**. A **regular** icosahedron has twenty **equilateral** triangles as faces and is a member of a family of solids sometimes referred to as the **Platonic solids**.

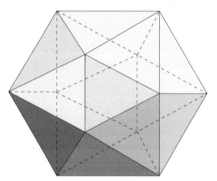

icosahedron: *a regular icosahedron*

image 1. The result of applying a **function** or **mapping** to an **element** of a **domain**. For example, if:

$$f(x) = x^2 - 5$$

for $x \in R$, then

$$f(3) = 3^2 - 5 = 4,$$

that is, if 3 is introduced into the function f then the image of 3 through f is 4. In this way each element of the domain of a function has precisely one image that is a member of the range of the function.
2. In **geometry** the 'image' is the result of some **transformation**.

imaginary part In a **complex number** of the form:

$$a + ib,$$

the imaginary part is b, which is the multiplier of $i = \sqrt{-1}$. For example, the imaginary part of $3 + 2i$ is 2.

Imperial system A system of measures fixed by the parliament of the U.K., which persisted until it was replaced by the **metric system** in 1963. See **unit conversions**.

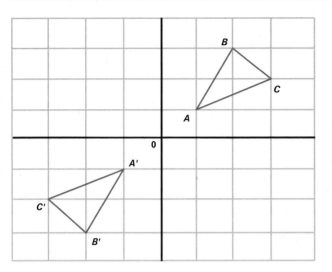

image: *triangle ABC is related to its image A'B'C' by reflection through the origin*

improper fraction See **fraction**.

incentre The point of **intersection** of the **angle bisectors** in a **polygon**.

inch A **unit** of length. See **unit conversions**.

incircle A circle drawn with its centre coinciding with the **incentre** of a triangle and its edges just touching the sides of the **triangle**. The incircle is sometimes called the **inscribed circle** of the triangle.

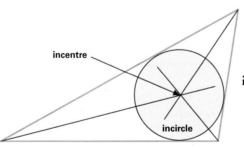

incircle: *the incircle of a triangle*

independent variable See **variable**. In an **equation** of the form $y = f(x)$, x is said to be the independent variable since an *independent* choice of x yields a value of y that is *dependent* on the choice of x.

independent event See **event**.

index (*pl.* **indices**) A numerical symbol representing the **power** to which a **base** is to be raised. For example, the index in 6^2 is 2.

When numbers are multiplied, divided or further raised by a power the following rules apply for their indices:

- $x^a \times x^b = x^{a+b}$ (add the indices when multiplying).
- $x^a \div x^b = x^{a-b}$ (subtract indices when dividing).
- $(x^a)^b = x^{ab}$.

The concept of zero, negative, and fractional indices may be understood through applying the above rules:

- $x^a \div x^a = x^0 = 1$.
- $1/x^a = 1 \div x^a = x^{-a}$.
- $\sqrt[n]{x} = x^{1/n}$.
- $\sqrt[b]{x^a} = x^{a/b}$.

For index notation see **standard form**.

inequality A mathematical statement of ordering in terms of size. For example:

$$5 > 2 \text{ (5 is greater than 2)},$$

$$2 < 7 \text{ (2 is less than 7)},$$

$$x \geq 5 \text{ (x is greater than or equal to 5)},$$

$$y \leq 10 \text{ (y is less than or equal to 10)}.$$

Inequalities remain true if the same quantity is added or subtracted from both sides, or if both sides are multiplied or divided by the same **positive number**. For example:

- Both sides of $5 > 3$ may be increased by 2 to yield $7 > 5$.
- Both sides of $5 > 3$ may be decreased by 2 to yield $3 > 1$.
- Both sides of $5 > 3$ may be multiplied by 2 to yield $10 > 6$.
- Both sides of $10 > 6$ may be divided by 2 to yield $5 > 3$.

However, multiplying or dividing by a **negative** number or inverting both

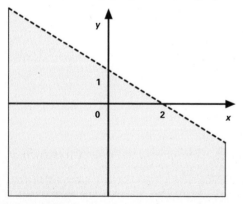

inequality: *the shaded area represents the inequality $x + 2y < 2$*

sides of the inequality changes the sense of the inequality. For example:
- Both sides of 5 > 3 may be multiplied by –2 to yield –10 < –6.
- Both sides of 10 > 6 may be divided by –2 to yield –5 < –3.
- Both sides of 5 > 3 may be inverted by to yield 1/5 < 1/3.

See also **inequation**.

inequation A mathematical statement involving **inequalities**. Inequations are easily illustrated graphically. Inequalities with two **variables** x and y, may have many solutions and all the values of x and y that satisfy the inequality may be found by drawing a **graph**. The graph may be drawn with a dotted boundary line to represent < or > and a solid boundary line to represent ≤ or ≥. For example, the inequation:

$$x + 2y < 2$$

has a boundary represented by a dotted line because of the < in the inequality (see diagram opposite).

infinite A quantity is said to be infinite if it is larger than any fixed limit. For example, the **set** of **integers** {1, 2, 3, 4, 5, 6, ... } has no fixed limit. However, the set of letters {a, b, c, d, ... z} is said to be *finite* since the members of the set can be counted (26 members). The symbol ∞ is used to represent an infinite quantity.

inflection A turning-point on a curve.
A stationary inflection is a point on a curve at which the **curve** changes from being concave upwards to **concave** downwards (or vice versa), whilst having a zero gradient at the point of inflection.

A *nonstationary inflection* is a point on a curve where the **curve** changes from being concave upwards to **concave** downwards (or vice versa) whilst having a nonzero gradient at the point of inflection. See illustration overleaf.

inscribed circle A circle drawn with its centre coinciding with the **incentre** of a **polygon** and its edges just touching the sides of the polygon. See **incircle**.

integer A **whole number**. For example:

... –5, –4, –3, –2, –1 , 0, +1, +2, +3, +4, +5...
negative integers   positive integers

intercept In the graphical representation of an **equation**, the distance from the origin to the point at which the line cuts the axis. The diagram shows x and y intercepts of a straight line. See illustration overleaf.

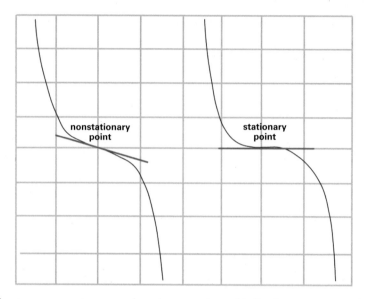

inflection: *nonstationary and stationary points of inflection on a curve. At a point of inflection, the tangent crosses over the curve.*

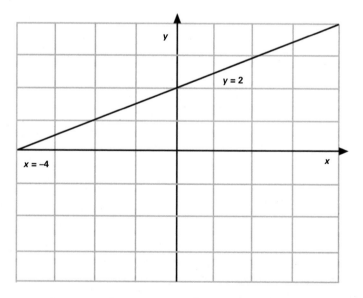

intercept: *the intercept on the x-axis is at x = –4 and the intercept on the y-axis is at y = 2*

interest A charge made on a loan or money received on a capital investment.

Compound interest is interest earned during a period calculated on the basis of the original **sum** together with interest earned from previous periods. If the compound interest is $x\%$, and the original investment is £Q, then the value of the investment after n years is:

$$£Q \times [(100 + x)/100]^n.$$

For example, the appreciation of £100 invested at 10% per year for 3 years may be calculated as follows:

	First Year	Second Year	Third Year
Sum	£100	£110	£121
Interest	£ 10	£ 11	£ 12.10
Value	£110	£121	£133.10

In the above table, the *Sum* is the amount on which the interest is calculated, the *Interest* is at 10% per year, and the *Value* is the value of the investment at the end of each year (Sum + Interest).

Simple interest (I) is calculated by multipying the amount invested (sometimes called the *principle*, P) by the length of time (T) the money is invested and the rate of interest (R) converted to a fraction; that is:

$$I = (P \times R \times T)/100.$$

For example, the simple interest if £5000 is invested at 7% over 4 years is:

$$I = (5000 \times 7 \times 4) \div 100 = £1400.$$

interior angle An angle between the sides at the **vertex** of a **polygon**. For interior angles between a transversal and two parallel lines, see **angle**.

interpolation The process of calculating the **value** of a **function** at a point by using known values of the function at either side of the point.

For example, in the diagram (see illustation overleaf) the value of A is interpolated from the data points B and C either side.

interquartile range The **difference** between the upper and lower **quartiles**; that is,

$$\text{interquartile range} = \text{upper quartile} - \text{lower quartile}.$$

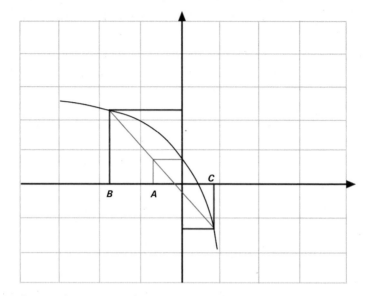

interpolation: *the position of A is interpolated from the data points B and C*

For example, consider the following **sample** of **data**:

8, 11, 12, 13, 15, 15, 18, 19.

The lower quartile occurs at 11.5, the upper quartile occurs at 16.5, so the interquartile range is:

16.5 − 11.5 = 5.

The **median** of this data occurs at 14.

intersection 1. Sets of points that are common to two or more lines or curves.
2. The intersection of two sets is the **set** of **elements** common to both sets. For example, the intersection of the two sets:

{2, 3, 4, 5, 6, 7} and
{2, 6, 9, 13, 45}

is the set {2, 6}.

interval A part of the **real number** line.
An *open interval* is one in which the 'end-point' values are included. For example, the interval [−2,3] includes the values −2 and 3; that is, [−2,3] is the **set** $\{x : -2 \le x \pounds 3\}$.
A *closed interval* is one in which the end-point values are not included.

For example, the interval $]{-}2,3[$ does not include the values -2 and 3; that is, $]{-}2,3[$ is the set $\{x : -2 < x < 3\}$.

A *mixed interval* is one in which only one of the end point values is included. For example, the interval $[-2,3[$ includes the value -2 but not the value 3; that is, $[-2,3[$ is the **set** $\{x : -2 \leq x < 3\}$.

invariant A quantity or property is said to be invariant under a transformation if the **transformation** leaves the quantity or property unchanged. For example, the form of a regular **hexagon** remains invariant if one applies a rotation transformation of $60°$ about its centre.

inverse A given mathematical **operation** may be reversed by the application of the corresponding *inverse operation*. For example, the operation of adding 7 to x may be reversed by applying the inverse operation of subtracting 7 from x; that is:

$$x + 7 - 7 = x.$$

The table below lists some simple **arithmetic** operations and their inverses. For **plane transformations** the inverse operation may be represented as an inverse **matrix**.

Operation	Inverse operation
+ (addition)	− (subtraction)
− (subtraction)	+ (addition)
× (multiplication)	÷ (division)
÷ (division)	× (multiplication)
x^n (x to the power n)	$\sqrt[n]{x}$ (the nth root of x)
$\sqrt[n]{x}$ (the nth root of x)	x^n (x to the power n)

inverse proportion If y is inversely proportional to x, then if x increases in size by a by a given factor, y will diminish in size by the same factor. One may represent this with the mathematical symbols $y \propto 1/x$. The diagram overleaf shows a graphical representation of a typical inverse proportion relationship.

inverse square law If y is related to x by an inverse square law, then if x increases in size by a given **factor**, y will diminish in size by the **square** of the same factor. One may represent this with the mathematical symbols $y \propto 1/x^2$.

The inverse square law appears quite frequently in physics. For example, the intensity of light from a point source diminishes as the inverse square of

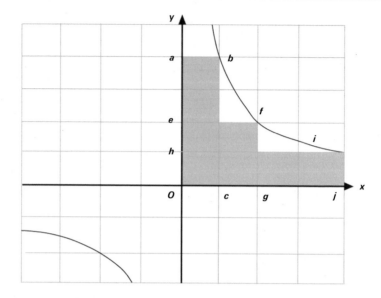

inverse proportion: *a graph of the equation yx = k, where k is a constant.*
The areas Oabc, Oefg, and Ohij are all equal to the constant.

the distance from the source; that is, if one moves to double the distance away from the source, the intensity diminishes by a factor of 2^2 or 4.

irrational number A number that is not **rational**; that is, a number that cannot be expressed as a **ratio** of two **integers**. For example, π, e, $\sqrt{2}$, and $\sqrt{7}$ are all irrational numbers. See also **rational number** and **real number**.

isosceles trapezium A **trapezium** in which the two nonparallel sides are of equal length.

isosceles triangle A **triangle** that has two sides equal (and unequal to a third). The **angles** opposite the equal sides are always equal.

iterative A relationship between consecutive terms of a sequence is said to be an iterative relationship. For example, the terms (t) in the **Fibonacci sequence** {1, 1, 2, 3, 5, 8, 13,...} may be defined by the iterative relationship:

$$t_{n+2} = t_{n+1} + t_n;\ t_1 = 1.$$

This expression indicates that each term is obtained by adding the two previous terms, with the first term equal to 1. The process of calculating these terms is called an iterative process. In such a process, input **data** is used to produce an output which is then used as an input in the next step of the iteration.

K

kilo- A prefix with **symbol** 'k' meaning 'one thousand times' (10^3). For example, a kilogram of potatoes (1 kg) is equivalent to one thousand grams in mass; i.e. 1000 g.

kilogram A **unit** of **mass**: 1 kilogram = 1000 grams. See **unit conversions** for a complete list of conversions from the **Imperial system** to the **metric system**.

kilometre A **unit** of distance: 1 kilometre = 1000 metres. See **unit conversions** for a complete list of conversions from the **Imperial system** to the **metric system**.

kite A **quadrilateral** having two pairs of adjacent sides **equal**. The diagonals on a kite are **perpendicular** but only one of them bisects the kite.

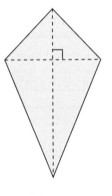

kite

knot A maritime unit of speed equivalent to one **nautical mile** an hour. See **unit conversions**.

L

latitude Lines of latitude on the Earth are imaginary circles that are drawn on the surface parallel to the **equator**. The latitude of a point P on the Earth is illustrated in the diagram below as the angle A. All latitude lines apart from the equator are **small circles**. See also **great circle**.

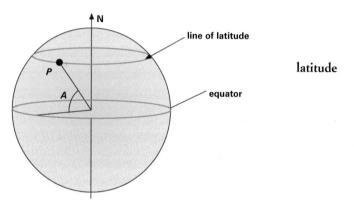

latitude

law A mathematical statement that is found to be true under certain conditions.

leading diagonal See **diagonal**.

limit **1**. The limit of a **sequence** is the number to which the terms of the sequence **converge**.
2. The limit of a **series** is the number to which the **sum** of the series tends.

line bisector See **bisect**.

line graph A **graph** constructed by joining a number of **points** together. These points are known **values** of a given **variable**. However points on **line segments** drawn between these known values may or may not have a meaning.

line of best fit If **points** on a **scatter graph** exhibit any **correlation**, a line of best fit may drawn through them. The line of best fit has about the same number of points above and below it. If there are only a small number of points, one can calculate the mean point and then position the line so that it passes the **mean point** with an equal number of the other points on either side of the line.

line of symmetry See **axis** and **symmetry**.

line segment A part of a line. For example, an **interval** may be thought of as a line segment of the **real number** line. The interval $-2 \le x \le 3$ is the line segment of the real number line which includes its end points -2 and 3. Just as with intervals, the end points of the line segment do not necessarily have to be part of the line segment.

linear If two **variables** may be related by a straight **line graph** then they are said to be related by a linear relationship. See **Appendix 1**.

linear programming A graphical method of finding the optimum solution to a problem that has a whole set of possible solutions.

 For example, the diagram on the next page shows how the quantity $L = x + y$ may be maximized subject to the constraints that $y - 2x \ge 2$, $x \ge 0$, and $7 \ge y \ge 0$. One may consider the construction of the diagram as a successive application of constraints:

(1) $L = x + y$ means that L may occupy the whole (x,y)-plane.
(2) The constraints $x \ge 0$ and $7 \ge y \ge 0$ mean that L must lie on or below $y = 7$, on or above $y = 0$, and on or to the right of $x = 0$.
(3) The constraint $y - 2x \ge 2$ means that $y \ge 2x + 2$, which means that y in L must lie on or above the line $y = 2x + 2$.

 The shaded area shows all the possible solutions. The maximum L corresponds to the point marked *max of L*.

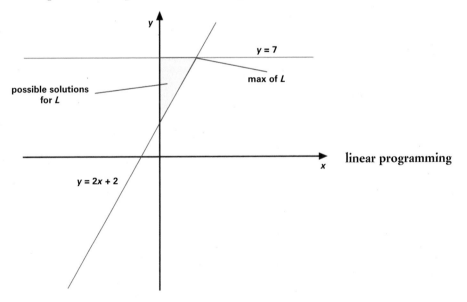

litre A measure of **volume** equal to 1000 cubic centimetres. 1 litre = 1000 cm³. See **unit conversions**.

locus (*pl.* **loci**) The path of a point that moves according to some rule. For example, the locus of a point that is a constant distance from a fixed point is a **circle**.

Graphical representations of loci are conveniently represented as **equations**. For example a circle may be referred to a set of **Cartesian coordinates**. The diagram overleaf shows a point on a circle as having the coordinates (x,y). With the restriction that the distance from the centre $(0,0)$ should always be r, the equation for the circle is easily derived from an application of **Pythagoras theorem**:

the square of the hypotenuse = the sum of the squares
of the other two sides

$$r^2 = x^2 + y^2.$$

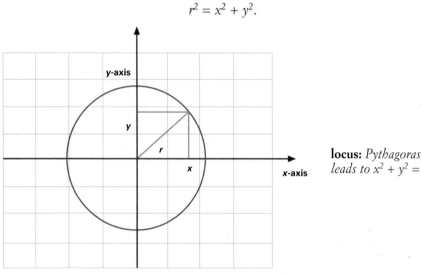

locus: *Pythagoras leads to* $x^2 + y^2 = r^2$

logarithm The logarithm of a number to a given **base** is the **power** to which the base has to be raised by to obtain the number. For example, $\log_{10}(100) = 2$ means that the base 10 must be raised by 2, i.e. squared, to get 100; that is $10^2 = 100$.

Logarithms to the base 10 are called **common logarithms** and to base **e** are called **natural logarithms**. The natural logarithm appears so frequently in physics and mathematics that it has its own symbol; $\log_e(x)$ may be written in a shorter form as $\ln(x)$.

longitude A line of longitude is an imaginary semicircle drawn on the surface of the Earth from the North pole to the South pole. The longitude of a point on the Earth is measured in **degrees**. It is the angular

displacement of the point with respect to the line of zero longitude which by convention is the Greenwich **meridian** (the line of longitude that passes through Greenwich in England).

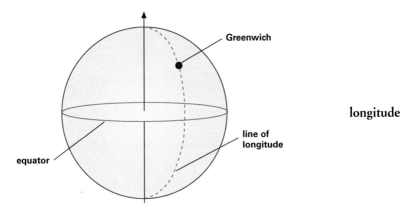

longitude

lower bound Any measurement to a nearest unit is actually referring to an **interval** of possible values centred on the measurement. The bottom end of the range of possible values is called the *lower bound*. For example, a measurement of 25 cm to the nearest cm could be a **value** in the range 24.5 cm to 25.5 cm. The lower value of 24.5 cm is the lower bound of this range. See also **approximation, accuracy, error,** and **upper bound.**

lower quartile See **quartile.**

lowest 1. The *lowest common multiple* (LCM) of two numbers is the smallest number that is a **multiple** of both the given numbers. For example, since

$$36 = 2 \times 2 \times 3 \times 3, \text{ and}$$

$$90 = 2 \times 3 \times 3 \times 5$$

the LCM of 36 and 90 is:

$$2 \times 2 \times 3 \times 3 \times 5 = 180.$$

2. The *lowest common* **denominator** of a set of fractions is the denominator that is the lowest common multiple of all the denominators. For example, the lowest common denominator of 3/4, 5/6 and 1/3 is the LCM of 4, 6, and 3; that is, 12.

3. A **fraction** which is in its simplest form is said to be written in its *lowest terms*. For example, the fraction 16/24 may be written in its lowest terms as 2/3.

M

magnitude 1. The length of a **vector**. When a vector is referred to a **Cartesian coordinate** system, the magnitude is easily calculated by an application of the **Pythagoras theorem**. The diagram shows a vector *A* represented as an arrow; the length of *A* is found by the application of Pythagoras to its **components**.
2. The magnitude of a **scalar** is its size without regard for **sign**.

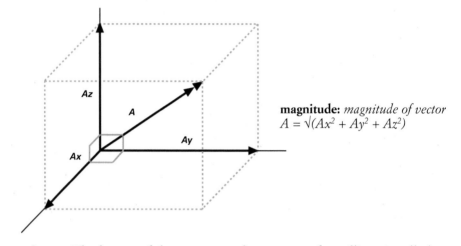

magnitude: *magnitude of vector*
$A = \sqrt{(Ax^2 + Ay^2 + Az^2)}$

major 1. The longer of the two **axes** of **symmetry** of an **ellipse** is called the major axis. It is the axis upon which the *foci* of the ellipse lie.
2. When a **sector** is demarkated in a **circle**, the largest sector is called the *major sector*. Similarly, the largest **arc** in such a demarcated circle is called the *major arc*. See **minor**.

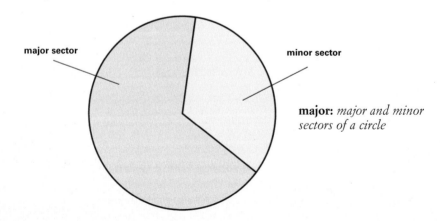

major sector

minor sector

major: *major and minor sectors of a circle*

mantissa The fractional part of a **logarithm**. For example:

$$\log_{10}(0.056) = -2\log_{10}(10) + \log_{10}(5.6)$$
$$= -2 + 0.7482$$

has a numerical value –1.2518 but is made up of an **integer part** (–2) and a fractional part (0.7482). It is this fractional part that is called the mantissa. See also **characteristic**.

mapping A **relation** that can be represented using a mapping diagram. The left-hand side of a *mapping diagram* is called the **domain**; the right-hand side of the diagram is called the **set** of **images**. Mappings can be **linear** or nonlinear depending on how the domain and the image set are related. For example, the illustration shows two mapping diagrams for the mapping $x \rightarrow x + 2$ (linear) and the mapping $x \rightarrow x^2$ (nonlinear).

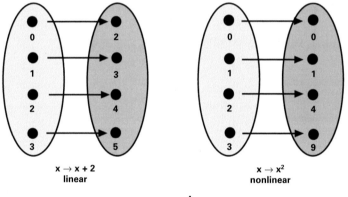

$x \rightarrow x + 2$
linear

$x \rightarrow x^2$
nonlinear

mapping

mass The amount of matter in a body is its mass. Mass is measured in **kilograms**.

matrix (*pl.* **matrices**) A rectangular array of numbers. Information may be stored in the rows and columns of a matrix; each position in the matrix contains a particular piece of information called an *element*.

A position **vector** is an example of a matrix of **order** 2 × 1; that is a matrix of two rows and one column. For example the position of a point P from the origin O may be written as:

$$\begin{pmatrix} x \\ y \end{pmatrix}$$

that is, the top number gives the horizontal distance from O to P and the bottom number gives the vertical distance from O to P.

Only matrices of the same order may be combined under **addition** or **subtraction** because the addition or subtraction must take place between corresponding positions in the matrices. The **multiplication** of a matrix by a **scalar** therefore just multiplies all the elements of the matrix by the scalar.

Matrices may be multiplied together but the rules of multiplication are much more complicated. One may **pre-multiply** a 2 × 1 matrix by a 2 × 2 matrix to leave a 2 × 1 matrix. For example:

$$\begin{pmatrix} a & c \\ b & d \end{pmatrix}\begin{pmatrix} x \\ y \end{pmatrix} = \begin{pmatrix} x' \\ y' \end{pmatrix}$$

is an expression of such a pre-multiplication. The values of x' and y' may be calculated using the following rules: First multiply along the top row of the matrix and down the column of the matrix, multiplying the first number in the row by the first number in the column and the second number in the row by the second number in the column and adding the results. The total calculation becomes:

$$\begin{pmatrix} a & c \\ b & d \end{pmatrix}\begin{pmatrix} x \\ y \end{pmatrix} = \begin{pmatrix} ax + by \\ cx + dy \end{pmatrix} = \begin{pmatrix} x' \\ y' \end{pmatrix}$$

Multiplication of matrices of other orders follows the above rules. The order of the result depends on the orders of the original matrices.

matrix: *multiplication of matrices*

Division may be defined for matrices but only if they are square. Division of a matrix A by a matrix B involves the multiplication of A by the *inverse* of B denoted B^{-1}. Square matrices with nonzero determinants have multiplicative inverses:

$$\begin{pmatrix} a & c \\ b & d \end{pmatrix} = \left(\frac{1}{D}\right)\begin{pmatrix} d & -b \\ -c & a \end{pmatrix}$$

where D is the **determinant** of the matrix. As one can see from this matrix equation one can only 'divide' by matrices with nonzero determinants. The multiplication of B with its inverse always results in the *identity matrix*:

$$(B)(B^{-1}) = (B^{-1})(B) = \begin{pmatrix} 1 & 0 \\ 0 & 1 \end{pmatrix}$$

matrix transformations See **transformations**.

maximum A point on a **graph** at its highest **value**. For example, the function in the diagram has a maximum at the point labelled *MAX*. **Tangent** lines are drawn to the curve before the maximum, on the maximum and after the maximum occurs. A simple way of defining a maximum is therefore obtained by considering the gradient properties of the curve; that is, a maximum is a point on a function where gradients of tangents to the curve change from **positive** through **zero** to **negative**. See **minimum**.

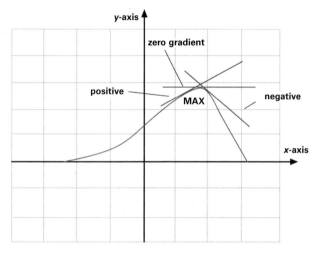

maximum: *the gradient of the tangent changes from positive through zero to negative on passing through a maximum*

mean The *arithmetic mean* value of a **sample** of **values** of a particular quantity is the **sum** of all the values divided by the number of values in the sample. The arithmetic mean is often called the **average** value although there are three different concepts associated with this word, namely, mean, **mode** and **median**.

   This concept of the mean is not really applicable to data delivered in the form of a **grouped frequency table**. Individual entries in the table refer to the number of values that belong to a particular **class interval**. With grouped **data** the mean value is constructed with the middle value of each class interval in the following manner. The mean value <x> is given by:

$$<x> = \frac{\sum_i x_i f_i}{\sum_i f_i}$$

where x_i and f_i are respectively the mid-value of the ith class interval and the frequency of the ith interval.

mean deviation See **deviation**.

mean point If there is a relatively small number of points in a **scatter graph**, a mean point may be calculated to aid in the construction of a **line of best fit**. For example, the scatter diagram illustrated shows a positive **correlation** between the **variables** A and B, and so the mean point may be used as a point of reference about which the prospective line of best fit may be rotated. A and B may be two positively correlated quantities, such as height and weight for a class of children.

A	2	2.5	3	4	5	5	6.5	7
B	1	3	2	3	3	5	5	6

The **means** of A and B are **respectively**:

$$\langle A \rangle = (2 + 2.5 + 3 + 4 + 5 + 5 + 6.5 + 7) \div 8,$$

= 4.4 to 2 **significant figures**,

$$\langle B \rangle = (1 + 3 + 2 + 3 + 3 + 5 + 5 + 6) \div 8$$

= 3.5 to 2 significant figures.
Therefore the mean point is (4.4, 3.5).

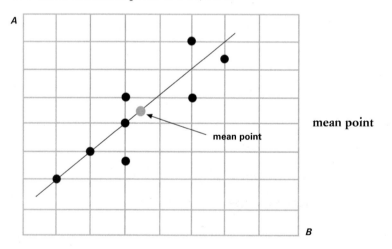

mean point

mean point

measure of dispersion A measure of the **dispersion** of a **sample** of **data**; that is, how spread out the data is from its **mean**. Possible measures of dispersion include mean **deviation**, standard deviation, **variance**, and **semi-interquartile range**. The **interquartile range** and **range** are also measures of spread but are concerned more with the size of the sample rather than how much the sample deviates from the mean.

median The middle **value** of a **sample** of **data** that is arranged in order. For example, the sample 3, 2, 6, 2, 2, 3, 7, 4 may be arranged in order as follows 2, 2, 2, 3, 3, 4, 6, 7. The median is the fourth value, which is 3. If there is an even number of values the median is the **mean** of the two middle values, for example, 2, 3, 6, 7, 8, 9, has a median of 6.5.

For data represented in a **grouped frequency table**, the median class may be defined. This is the **class interval** that contains the median value. If the data is then represented in the form of an **ogive** or **cumulative frequency** graph, the median value is just the value that corresponds to the mid-point of the frequency **range**.

For example, in the diagram the data ogive represents the number of chocolate bars that are consumed per week by children in a sample of 30 children. The median number that a child eats per week is read off as the number corresponding to the median cumulative frequency, which is 3.

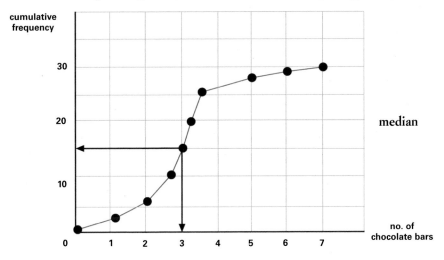

mega- A prefix with **symbol** 'M' meaning 'one million times'(10^6). For example, a megavolt (MV) is one million volts.

member A number, letter, **symbol**, etc., that belongs to a **set**. To make mathematical expressions more concise, the symbol $\in$ is used to mean '*is a member of*', for example:

$$2 \in \{\text{even numbers}\},$$
$$6 \notin \{\text{prime numbers}\},$$

the first expression says that '2 is a member of the set of **even** numbers', the second expression says that '6 is not a member ($\notin$) of the set of **prime** numbers'.

mensuration The study of the mathematical methods employed to measure geometrical quantities such as lengths, **areas**, and **volumes**.

meridian A **great circle** drawn on the Earth's surface that passes through the North and South poles. All lines of **longitude** are parts of meridians. See **longitude**.

metre A **unit** of length. See **unit conversion**.

metric system A system of **units** of measurement. Metric units are graduated into factors of ten. For example, 1 **kilogram** = 1000 **grams**, 1 **metre** = 100 **centimetres**. See unit **conversion**.

micro- A prefix with **symbol** 'μ' meaning 'one millionth of' (10^{-6}). For example, 1 μm is 1 millionth of a metre.

mile A **unit** of length. 1 mile = 5280 **feet** or 1760 **yards**. See **unit conversion**.

milli- A prefix with **symbol** 'm' meaning 'one-thousandth of'. For example, a millimetre (1 mm) is one-thousandth of a metre.

million One thousand thousand (10^6).

minimum A point on a **graph** at its lowest **value**. For example, the function in the diagram has a minimum at the point labelled *MIN*.

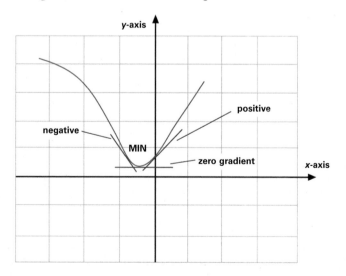

minimum: *at the minimum, the gradient of the tangent to the curve changes from negative through zero to positive*

Tangent lines are drawn to the **curve** before the minimum, on the minimum and after the minimum occurs. A simple way of defining a minimum is therefore obtained by considering the **gradient** properties; that is, a minimum is a point on a function where gradients of tangents change from **negative** through **zero** to **positive**. See **maximum**.

minor 1. The shorter of the two **axes** of **symmetry** of an **ellipse** is called the minor axis.
2. When one demarcates a **sector** in a **circle**, the smallest sector is called the *minor sector*. Similarly, the smallest **arc** in such a demarcated circle is called the *minor arc*. See **major**.

minute 1. A **unit** of time. There are 60 **seconds** in a minute and 60 minutes in an **hour**.
2. A measure of **arc** or **angle**. There are 60 minutes in one **degree**; that is, $60' = 1°$.

mixed number The expression of an improper **fraction** as the **sum** of an **integer** or **whole number** and a **fraction**. For example,

$$5/2 = 2\tfrac{1}{2},\ 17/3 = 5\tfrac{2}{3},\ -51/8 = -6\tfrac{3}{8}.$$

mode The **value** that occurs most often in a **sample**. For example, the mode of the sample: 2, 2, 3, 3, 3, 3, 4, 5, 5, is 3.
For data represented in a **grouped frequency table**, the *modal class* may be defined. This is the **class interval** which has the **highest frequency**. The mid-point of the modal class gives an approximate value for the mode. For example, the table below shows the weights of a class of children:

weight/kg	26–30	31–35	36–40	41–45
frequency	1	5	3	2

The modal class is 31–35 kg.

modulus 1. The size of a **real number** without regard for **sign**. The modulus of a real number x is denoted $|x|$. For example, $|-8| = 8$, $|3| = 3$.
2. **Complex numbers** may be represented on a diagram called an *Argand diagram*. The part of the complex number that is just a real number is represented on a **horizontal axis** (real axis). The part of the complex number that is a multiple of i is represented on a vertical axis (**imaginary axis**). The distance of the point corresponding to the complex number from the origin is the *modulus* of the complex number. For example, the diagram overleaf shows an Argand representation of a complex number

$z = a + ib$. The modulus of this number is $\sqrt{(a^2 + b^2)}$ and is denoted by $|a + ib|$.

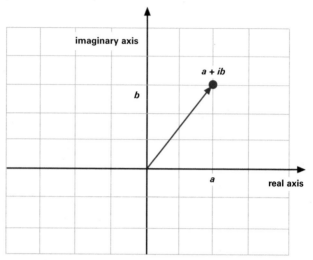

modulus: *the Argand diagram for a complex number*

multiple The multiples of a number are found by the **multiplication** of the number by each of 1, 2, 3, ... For example, the multiples of 10 are 10, 20, 30, 40, etc.

multiplication A basic **operation** of **arithmetic**. Multiplication is associated with the repeated application of the **addition** operation. For example:

$$4 \times 8 = 8 + 8 + 8 + 8 = 32.$$

Multiplication of numbers is **commutative**. However, multiplication in more abstract systems of mathematics can be less straightforward. For example, multiplication as an operation has a slightly different definition when the entities to be multiplied are **matrices**. In this case the operation is not commutative; that is, in general:

$$A \times B \neq B \times A,$$

where A and B are matrices.

mutually exclusive events See **event** and **probability**.

N

nano- A prefix with **symbol** 'n' meaning 'one thousand millionth of' (10^{-9}). For example, 1 nm is one thousand millionth of a metre.

natural numbers The set of positive **integers** or **whole numbers** (counting numbers). The **set** of natural numbers

$$\{1, 2, 3, 4, 5, \ldots\}$$

is often denoted by N.

natural logarithm A **logarithm** to the **base** of **e**.

nautical mile The distance equivalent to one **minute** of **arc** measured on a **great circle** on the Earth's surface. It is equal to 6080 **feet**. Hence 60 nautical miles of arc along a great circle **subtends** an angle of 1° at the centre of the Earth. See **unit conversions**.

negative Describing a number whose **value** is less than **zero**. For example, –2, –4, –7.5, are all negative numbers. See also **positive**.

net A two-dimensional representation of the surface of a three-dimensional **solid**. For example, in the illustration the net of a cube and its corresponding solid show the correspondence between the **faces** of the cube and the sections of the net. One may construct the solid cube by folding the net along the dotted lines.

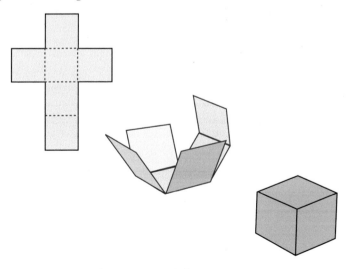

net

network A system of **points** (called **nodes**) connected together by **arcs**. A network divides the **plane** into a number of regions. There is a relationship between the number of nodes, arcs and regions which is encompassed in Euler's formula. See **Appendix 3**.

newton A **unit** of force. An unbalanced force of 1 newton accelerates a mass of 1 kg by 1 m s^{-2}.

node A **point** in a **network** to which one or more **arcs** lead.

nonagon A **polygon** with nine **sides**. A **regular** nonagon has all its sides of equal length, and each of its interior **angles** measures 140°.

normal **1**. A *normal* at a point on a curve may be constructed by first constructing a tangent to the curve at the point. The normal is then the line that is **perpendicular** to this tangent through the point.
2. A line or **vector** is said to be normal to a **plane** if it is perpendicular to all lines belonging to the plane.

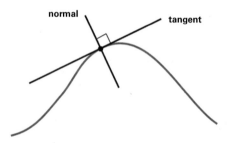

normal: *the normal to a curve*

normal distribution See **distribution**.

notation A system of signs or **symbols** together with the conventions regarding their use. For example, the mathematical notation:

$$\{z: 2 < z < 5\} \subset \{t: 4 < z + 2\},$$

corresponds to the statement 'the **set** of z **values** from the open **interval** between 2 and 5 is a **subset** of the set of t values whose values come from a range that satisfies the inequality $4 < z + 2$'. The notation is therefore a very concise way of writing down a lot of mathematical information.

nought The **numeral** 0 or **zero**.

null set The **empty set** denoted by the symbol Ø. The null set is a set of no members, for example, the set of odd numbers that are also factors of 2 is a null set.

numeral A **symbol** used to denote numbers. For example:

$$0, 1, 2, 3, 4, 5, 6, 7, 8, 9$$

are numerals of Arabic origin.

$$I, V, X, L, C, D, M$$

are numerals of Roman origin.

numerator The number above the line in a **fraction**. See also **denominator**.

O

object In mathematics, one applies a **transformation** to an object to form an **image**. See **enlargement** for an example of this.

observation sheet A form designed for the recording of **data**. The design of this form may be crucial if the speed of taking data down is important. For example, a well-designed observation sheet may ease the recording of data in a traffic **survey**.

obtuse A adjective used to describe an **angle** that is greater than 90° but less than 180°.

octagon A **polygon** that has eight **sides**. A **regular** octagon has all its sides of equal length, and each of its interior **angles** measures 135°.

octahedron A **polyhedron** that has eight **faces**. A regular octahedron has eight **equilateral triangles** as faces and is a member of a family of solids sometimes referred to as **Platonic solids**.

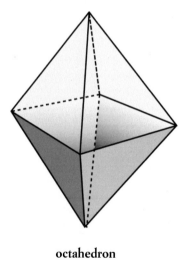

octahedron

odd number A number that is not a **multiple** of 2. A general **formula** for an odd number may be written in terms of an arbitrary whole **number**: $x = 2n + 1$; x is always odd if n is a whole number.

ogive An alternative term for a **cumulative frequency** graph.

operation A rule for combining **members** of a **set**. Operations in arithmetic are often described as **binary** as they combine two members at a time to produce another member of the set. For example, the **addition** operation acting on the set of **natural numbers** (N) leads to expressions of the form $a + b = c$, where a, b, and c all belong to the set N.

opposite side The side in a **right-angled triangle** opposite to the **angle** under consideration.

order **1.** The shape of a **matrix** in terms of the number of rows and columns. A matrix of order $n \times m$ is a matrix with n rows and m columns.
2. The number of **arcs** that lead to a given node in a **network**.
3. A figure that maps to itself under **reflections** about n lines is said to have line **symmetry** of *order n*.
4. A figure that maps to itself under n **rotations** about an **axis** is said to have rotational symmetry of order n *about* the axis.

order of magnitude A numerical solution to a problem is said to have the right order of magnitude if its size is reasonable compared to an estimated solution.
    For example, a student may miscalculate the expression $256 \div 45$ and get an answer of 56.9 to three **significant figures**. The teacher would hopefully spot that this could not be the case since an estimate, $250 \div 50$, leads to an answer of 5. The student's answer is of the wrong order of magnitude and the answer to $256 \div 45$ should be 5.69 to three significant figures.

ordered pair A pair of numbers whose order bears some significance; i.e. a, b has a different meaning to b, a.
    For example, the coordinate of a point in a two-dimensional **Cartesian coordinate** system is an ordered pair. The ordered pair (a, b) denotes the point in the Cartesian plane that is a units from the origin along the **abscissa** and b units from the origin along the **ordinate**.

ordinate The y-**coordinate** of a point referred to a **Cartesian coordinate** system.
    For example, in the diagram the two points P and Q have ordinates of 3 and -2 respectively. See illustration overleaf.

origin The point where the **abscissa** and **ordinate axes** cross in a coordinate system. It has the coordinates $(0, 0)$ and the coordinates of all other points are measured relative to the origin.

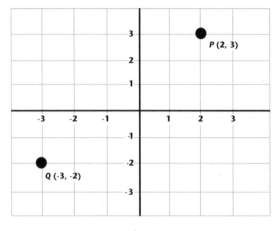

ordinate

orthocentre The point at which the **altitudes** meet in a **triangle**.

oscillate A **series** is said to oscillate if its **partial sums** oscillate. For example, $1 - 2 + 3 - 4 + 5 - \ldots\ldots$ has partial sums:

$1 = 1,$
$1 - 2 = -1,$
$1 - 2 + 3 = 2,$
$1 - 2 + 3 - 4 = -2,$
$1 - 2 + 3 - 4 + 5 = 3,$
etc.

Since these oscillate in value:

$$1, -1, 2, -2, 3, \ldots ,$$

the series itself is said to oscillate.

ounce A **unit** of weight. There are 16 ounces in 1 pound (16 oz = 1 lb). See **unit conversion**.

outcome The result of a statistical **trial** or other activity involving uncertainty.

P

parabola The **curve** that is the result of plotting a **quadratic equation**. The general quadratic:

$$y = ax^2 + bx + c$$

is a curve whose form depends on the constants a, b, and c. For example, in situations where the **common formula** may be used, the **roots** of the equation; that is, the points of **intersection** with the **x-axis** are given by:

$$x_0 = [-b \pm \div\sqrt{(b^2 - 4ac)}]/2a.$$

The intersection of the curve with the y-axis is easily found by considering the equation for the value $x = 0$; this leads to a y intersection of c. See **Appendix 1**.

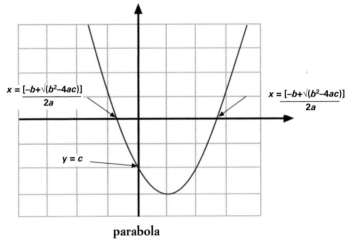

parabola

parallel Describing lines that when extended in the same direction are equidistant in all parts. Thus, no matter how far they are extended they always remain the same distance apart.

parallelepiped A **polyhedron** whose **faces** are all **parallelograms**.

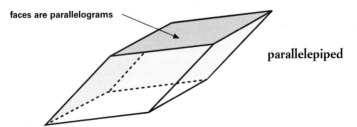

parallelogram A **quadrilateral** whose opposite sides are equal and **parallel**. All parallelograms have the following properties:
(1) The diagonals **bisect** each other and also bisect the whole parallelogram.
(2) The opposite **angles** are equal.
(3) The parallelogram has a rotational **symmetry** of **order** 1.

parallelogram rule A method of obtaining the **resultant** of two **vectors** by placing them so that they form two adjacent sides of a **parallelogram**. The resultant is then the diagonal of the parallelogram.

parameter A quantity in an analysis or experiment that affects the **values** of other quantities when it is varied.
    For example, the characteristics of a **parabola** given by the **quadratic**:

$$y = ax^2 + bx + c,$$

are specified by the parameters a, b, and c. Where the **common formula** may be used, the intersections of the curve with the x-axis occur at:

$$x = [-b \pm \sqrt{(b^2 - 4ac)}]/2a.$$

The intersection with the y-axis occurs at $y = c$, and the **turning point** occurs at:

$$x = -b/2a \text{ and } y = (-b^2/4a) + c.$$

Pascal's triangle A pattern of numbers derived from a triangular structure; each number is the **sum** of two numbers directly above it.
    The numbers in each row of Pascal's triangle are the **binomial coefficients** occurring in the expansion of $(x + y)_n$ for various values of n.

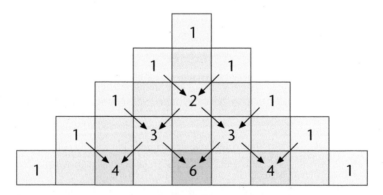

Pascal's triangle

pentagon A **polygon** that has five **sides**. A **regular** pentagon has all its sides of equal length, and each of its interior **angles** measures 108°.

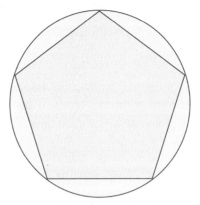

pentagon: *a regular pentagon inscribed in a circle*

percentage The expression of a **fraction** in terms of a whole that has 100 parts. For example, a **quarter** (1/4) expressed as a percentage is just '25 parts in 100' or 25%. To express an amount *a* as a percentage (a)% of a whole *b* one applies the following formula:

$$a\% = (a \div b) \times 100\%.$$

Percentage changes. A quantity may change over time and a convenient way of expressing this change is in terms of a percentage of the original amount. For example, if a quantity has an original **value** of x_0 and changes to a value *x* then the percentage change is given by:

$$(|x - x_0| \div x_0) \times 100\%,$$

where the vertical lines denote the magnitude of $(x_0 - x)$.

Percentage uncertainty. When a measurement is made with an instrument, it only really has a meaning to an **accuracy** of the smallest increment measurable by the instrument. For example, a **metre** rule with a smallest increment of 1 mm may be used to measure a length of 174 mm. The percentage uncertainty in the measurement is then:

$$\pm(1/174) \times 100 = \pm0.57\%.$$

percentile One of 100 parts into which the **range** of a **frequency distribution** may be divided. For example, 20% of the data lies below the 20th percentile. The **median** value of a distribution occurs at the fiftieth percentile.

perfect number A number that is equal to the **sum** of its **factors** (excluding the number itself). For example:

$$6 = 1 + 2 + 3,$$
$$28 = 1 + 2 + 4 + 7 + 14.$$

The third perfect number is 496.

perimeter The distance measured around the **boundary** of a figure. For the regular **polygons**, the total perimeter tends to $2\pi r$ as the number of sides increases, where r is the **radius** of the **circle** that just touches the **vertices** of the polygon. In the diagram on the next page, the **hexagon** is drawn within a circle of radius r. If one imagines the number of sides increasing from just six to a very large number, the angles **subtended** at the centre get smaller and smaller and the sides get shorter and shorter.

From the diagram, the perimeter P for a polygon of n sides is:

$$P = 2nr\sin\left(\frac{\pi}{n}\right)$$

For $n \to \infty$:

$$P = \lim_{n\to\infty}\left(2nr\sin\left(\frac{\pi}{n}\right)\right) = 2nr \times \frac{\pi}{n} = 2\pi r$$

which is the standard **formula** for the **circumference** of a circle of radius r, which is the limit of the perimeter for polygons of n sides as $n \to \infty$.

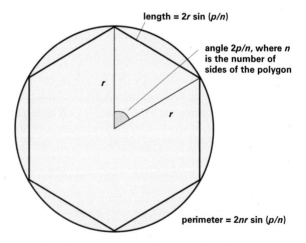

length = 2*r* sin (*p/n*)

angle 2*p/n*, where *n* is the number of sides of the polygon

r

r

perimeter = 2*nr* sin (*p/n*)

perimeter: *a hexagon inscribed in a circle*

periodic function A **function** that repeats regularly. Functions used in trigonometry are periodic, the *period* of the trigonometric function depending on the **factor** that the variable is multiplied by in the function. For example, in $y = \sin(2x)$ the factor is 2 and in $y = \sin(x)$ it is 1. These two functions have different periods as shown in the table:

x	$\sin(x)$	$\sin(2x)$
0°	0	0
45°	0.707	1
90°	1	0
135°	0.707	−1
180°	0	0
225°	−0.707	1
270°	−1	0
315°	−0.707	−1
360°	0	0

From the above table, it is evident that $\sin(x)$ has a period of 360° and $\sin(2x)$ has a period that is just 180°. See also **Appendix 1**.

permutation The ordered arrangement of a **set** of objects. For example, the ways of *permuting* three numbers {1, 2, 3} taken one at a time are as follows:

$$\{1, 2, 3\}, \{1, 3, 2\}, \{2, 1, 3\}, \{2, 3, 1\}, \{3, 1, 2\}, \{3, 2, 1\}.$$

The first **digit** may be chosen from 3 different numbers. The next choice is from 2 numbers and the final number can be chosen only from the 1 left. Therefore the number of permutations is 6, which may be written as 3! or 3 **factorial**. In general the number of permutations of n objects taken one at a time is $n!$ or n factorial.

For the permutations of n objects taken r at a time there are:

$$n!/(n - r)!$$

which is the number of combinations of r objects that can be chosen from a set of n, multiplied by the $r!$ ways of arranging them; that is, $r!C_r^n$ (see **combination**).

perpendicular At **right angles**. For example, the **Cartesian axes** for the x, y, and z directions are said to be *mutually perpendicular*.

Lines that are said to be perpendicular to a **plane** are also known as **normals** to the plane.

pi The Greek letter (π) which denotes the **ratio** of the **circumference** (C) of a **circle** to its **diameter** (D); that is $\pi = C/D$.

π is an **irrational** number: 3.141 592 653 5.... One commonly uses 3.14 or 22/7 as approximate values for π.

π is used in the **mensuration** of circular figures and **solids**. For example:
(1) The **area** of a circle of **radius** r is πr^2.
(2) The **volume** of a **cylinder** of radius r and height h is $\pi r^2 h$.
(3) The surface area of a **sphere** of radius r is $4\pi r^2$.

pictogram A pictorial method of representing **data** on a **graph**. A very simple example of a pictogram is shown overleaf, in which the profits of a shop keeper over a three-month period are represented by 'bags of money', each bag being equivalent to £100.

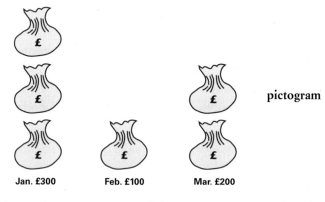

pictogram

Jan. £300   Feb. £100   Mar. £200

pie chart A representation of **data** as proportions of a whole 'pie' or **circle**. The circle is divided into sections. The number of **degrees** in the **angle** at the centre of each section represents the **frequency**. For example, the outcomes of a season of fixtures for a football team may be represented as shown in the diagram .

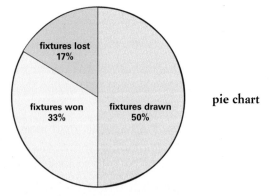

pie chart

pint A **unit** of **volume**. There are 8 pints in a **gallon**. See **unit conversions**.

plan The representation of anything projected onto a **plane** or flat surface. For example, the plan view of a building may be represented as floor plans or the **projection** of each floor onto the horizontal **plane**.

plane A flat surface. In **Cartesian coordinates** the equation of a plane is of the form:

$$ax + by + cz = d,$$

where a, b, c, and d are **constant parameters** and x, y, and z are the **Cartesian variables**.

plane of symmetry A plane that divides a **solid** into two **congruent** solids.

planning network A **network** constructed to illustrate the planning of a project. Networks of this type are very useful to visualize the application of **critical path** methods.

Platonic solids The **set** of five **regular polyhedra**. They are the **tetrahedron**, the **cube**, the **octahedron**, the **dodecahedron**, and the **icosahedron**.

These five polyhedra exhibit a *duality* amongst themselves that is based on a **face–vertex** interchange. The tetrahedron is seen to be *self-dual* and the cube and octahedron are **duals** of each other.

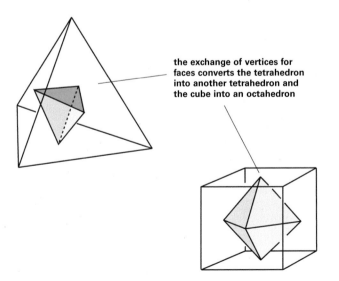

the exchange of vertices for faces converts the tetrahedron into another tetrahedron and the cube into an octahedron

Platonic solids

point A location in space.
Points in space are often described with reference to a **coordinate** system. Since a point designates location in space it is said to be one-dimensional and to occupy no **volume**; that is, a point has position but no real size.

polar coordinate system A system of **coordinates** specifying the position of a **point** with respect to an **origin** by the specification of its radial distance (r) from the origin and the **angle** (θ) made between the **radius vector** and a fixed line. Polar coordinates are of the form (r,θ) as opposed to (x,y) of the **Cartesian coordinate** system. The two systems are related by the transformation:

$$x = r \cos\theta,$$
$$y = r \sin\theta,$$
$$q = \tan^{-1}(y/x),$$
$$r = \sqrt{(x^2 + y^2)}.$$

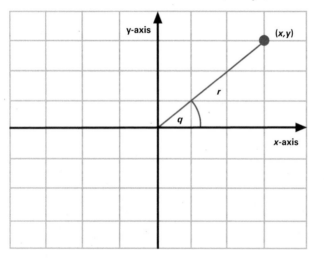

polar coordinate system

polygon A **plane** figure bounded by three or more sides. In general a polygon having n sides will have n interior **angles** that **sum** to $(n - 2) \times 180°$. **Regular** polygons have all sides of **equal** length and all their angles equal. The table opposite gives the properties of nine polygons.

polyhedron (*pl.* **polyhedra**) A **solid** figure bounded by four or more **faces** that are themselves **polygons**.
    A regular polyhedron has **regular** polygons as faces. There are only five regular polyhedra (tetrahedron, cube, octahedron, dodecahedron, and icosahedron). These regular polyhedra make up the **set** that is often called the **Platonic solids**.

Name of polygon	Number of sides	Sum of interior angles
Triangle	3	180°
Quadrilateral	4	360°
Pentagon	5	540°
Hexagon	6	720°
Heptagon	7	900°
Octagon	8	1080°
Nonagon	9	1260°
Decagon	10	1440°
Icosagon	20	3240°

polynomial An algebraic expression containing only **positive** powers of one or more **variables** x, y, z, …. For example:
(1) $ax^2 + bx + c$ is a general **quadratic** polynomial in x,
(2) $ax^3 + bx^2 + cx + d$ is a general **cubic** polynomial in x,
(3) $ax^2y^2 + bxy^3$ is a polynomial in x and y of **degree** 4,
where a, b, and c are the **coefficients** of the individual terms.

population In **statistics**, a collection or group of objects that are being studied. See also **sample**.

position vector See **vector**.

positive Describing a number whose **value** is greater than **zero**; that is, numbers that are found on the right of zero on a **directed** number line.

pound
1. A **unit** of currency. £1 is equal to 100 pence.
2. A unit of weight. There are 16 **ounces** to 1 pound (16 oz = 1 lb). See **unit conversions**.

power The **product** obtained by multiplying a number by itself a specified number of times. For example, the expression

$$2 \times 2 \times 2 \times 2 \times 2 \times 2 = 2^6$$

which is read as '2 to the power 6'. See also **square** and **cube**.

pre-multiply To perform a multiplication on a **mathematical** entity (number, matrix, etc.) in a particular order with the multiplication **factor**

appearing before the entity. With numbers this makes no difference because, for numbers, multiplication is **commutative**. However, other objects, such as matrices, do not share this property and the order of multiplication is important.

prime number A number that only has two factors, 1 and the number itself. There are an infinite number of prime numbers. For example, 2, 3, 5, 7, 11 are all examples of prime numbers. Note, 2 is the only **even** prime number.

prism A polyhedron that has a uniform cross-section along one of its lengths. The volume of a prism is therefore easily calculated by multiplying the area of the uniform cross-section by the length.

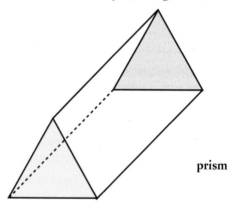

prism

probability The measure of the possibility of an **event** is its **probability**, p. For example, for N **equally likely events**, p may be expressed mathematically as the ratio of the number of events favourable to the outcome (n) to the total number of possible outcomes (N), i.e. $p = n/N$.
(1) *Exhaustive events* account for all possible outcomes. If a set of events is exhaustive then it is certain that one of them will occur; i.e. the **sum** of all the probabilities of the individual events will equate to 1.
(2) *Dependent events* lead to conditional probabilities because the occurrence of one event is conditional on whether or not the other event has occurred.
(3) *Independent events* are events for which the occurrence of one event has no influence on the occurrence of the other(s).
(4) *Mutually exclusive events* are events which cannot happen at the same time. For example, the toss of a coin can result either in a 'Head' or a 'Tail' but not both.
See also **combined probabilities** and **conditional probabilities**.

product The result of multiplying two or more entities together.

The rule for constructing the product of two matrices is more involved than just a **multiplication** of **elements**. See **matrix**.

progression **1.** *Arithmetic progression.* A **sequence** whose terms are obtained by the successive **addition** of a **common difference** to a first term. For example:

$$1, 3, 5, 7, 9, 11, \ldots,$$

is an example of an arithmetic progression for which the first term is 1 and the common difference is 2.

2. *Geometric progression.* A sequence whose terms are obtained by the successive **multiplication** of a **common ratio** to a first term. For example, the sequence:

$$1, 2, 4, 8, 16, \ldots,$$

is an example of a geometric progression for which the first term is 1 and the common ratio is 2.

projection The projection of a line onto a **plane** (or line) is formed by taking lines perpendicular to the plane (or line) from points on the given line (see illustration on next page). The projection of a **vector** onto the coordinate axes gives the **components** of the vector.

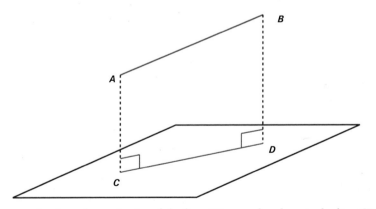

projection: *the projectionof the line AB onto the plane is the line CD*

proof A logical argument used to establish the truth of a mathematical statement. See also **hypothesis**, **theorem**, and **Appendix 3**.

proper fraction A **fraction** in which the **numerator** is less than the **denominator**. For example, the fraction 2/3 is a proper fraction and the fraction 3/2 is not.

proper subset A **subset** of a **set** that does not contain all the **elements** of the set. For example, if B is a proper subset of A then:

$$B \cap A = B,$$

that is, the elements that are common to both B and A are only in B. For example, the set {1,2,3,4,5} has a proper subset {1,2,3}. In this example the set {1,2,3,4,5} is not a proper subset of {1,2,3,4,5} because it contains all elements of A.

proportionality
1. *Direct proportionality.* Two quantities x and y are said to be related by a direct proportionality if an increase in x by a **factor** will lead to an increase in y by exactly the same factor. This can be represented by $x \propto y$, where $\propto$ means 'is proportional to'. For example, the **equation** $y = 3x$ describes a direct proportionality between x and y; that is, if the **value** of x is doubled from 2 to 4 then the value of y is doubled from 6 to 12.
2. *Inverse proportionality.* Two quantities x and y are said to be related by an **inverse** proportionality if an increase in x by a factor will lead to a decrease in y by exactly the same factor. For example, the equation $y = 1/x$ describes an inverse proportionality between x and y; that is, if the value of x is doubled from 2 to 4 then the value of y will be halved from ½ to ¼. See also **reciprocal**, **Appendix 1**, and **Appendix 2**.

protractor An instrument used for measuring **angles**.

pyramid A **polyhedron** on a triangular (see **triangle**), **square**, or polygonal (see **polygon**) **base**, with triangular **faces** meeting at a **vertex**.
     The **volume** V of a pyramid of base area A and perpendicular height h, is given by the formula:

$$V = \tfrac{1}{3}A \times h.$$

Pythagoras' theorem A **theorem** named after its discoverer Pythagoras, who was a Greek philosopher and mathematician (c.580–500 BC).
     The theorem states that in any **right-angled** triangle, the area of the square on the **hypotenuse** is equal to the **sum** of the areas of the squares on the other sides. See illustration opposite.

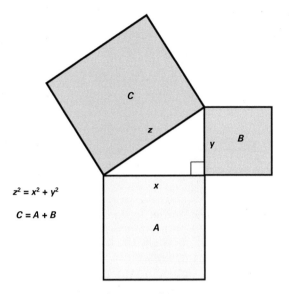

$z^2 = x^2 + y^2$

$C = A + B$

Pythagoras' theorem

Q

quadrant A **quarter** of a **circle**.
The quadrants of a circle may be used to illustrate the properties of the three **functions** used in trigonometry: **sine**, **cosine**, and **tangent**. In the diagram the point P with coordinates (x, y) moves on a circle with centre O and **radius** 1 unit. OP makes an **angle** θ with the **positive** x-axis. The angle increases as P **rotates** anticlockwise. For any angle θ, positive or negative, the sine and cosine of θ are

$$\sin\theta = y,$$
$$\cos\theta = x,$$
$$\tan\theta = y/x,$$

given by the coordinates of P (because the **hypotenuse** = 1). The signs of these trigonometric functions in the four quadrants of the circle depend therefore on the **signs** of x and y. So in the first quadrant all are positive, in the second quadrant only sine is positive, in the third quadrant only tangent is positive, and finally in the fourth quadrant only cosine is positive.

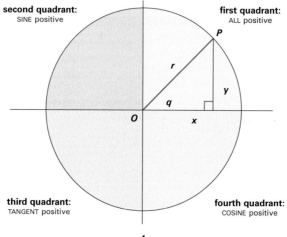

quadrant

quadratic A **polynomial** in which the highest **power** is 2. A general **equation** for a quadratic is as follows:

$$ax^2 + bx + c = y,$$

where a, b, and c are **constants**. When $y = 0$ the equation has two **solutions**, which are in general given by the **common formula**.

The **graph** of a quadratic **function** is called a **parabola**.
See **quartic**, **quintic**, and **Appendix 1**.

quadrilateral A **polygon** that has four sides. The **square, rhombus, rectangle, parallelogram, kite,** and **trapezium** are all special kinds of quadrilaterals. See **cyclic quadrilateral**.

quarter One of four equal parts into which an object may be separated. A quarter is represented as the **fraction** ¼.

quartic A **polynomial** in which the highest **power** is 4. A general equation for a quartic is as follows:

$$ax^4 + bx^3 + cx^2 + dx + e = y,$$

where a, b, c, d, and e are **constants**. When $y = 0$ the equation has four **solutions**. See **quadratic**, **quintic**, and **Appendix 1**.

quartile When the **range** of a **distribution** is divided into four parts, the **values** of the **variable** that correspond to the boundaries are called quartiles. The **value** one quarter of the way from the lower end of the range is the *lower quartile*. The middle value of the range (the *second quartile*) is the **median**. The value three-quarters of the way from the lower end of the range is called the *upper quartile*. For example, a calculation of the **interquartile range** demonstrates this idea:

$$interquartile\ range =$$
$$upper\ quartile - lower\ quartile.$$

Consider the following **sample** of **data**:

$$8, 11, 12, 13, 15, 15, 18, 19.$$

The lower quartile occurs at 11.5, the upper quartile occurs at 16.5, so the interquartile range is:

$$16.5 - 11.5 = 5;$$

the **median** of this data occurs at 14.

quintic A **polynomial** in which the highest **power** is 5. A general equation for a quintic is as follows:

$$ax^5 + bx^4 + cx^3 + dx^2 + ex + f = y,$$

where a, b, c, d, e, and f are **constants**. When $y = 0$ the equation has five solutions. See **quadratic**, **quartic**, and **Appendix 1**.

quotient The number of times one quantity is contained in another. For example, the division of 2736 by 82 is a quotient of 33 with a remainder 30; that is

$$2736/82 = 33 + (30/82).$$

quota sampling A method of taking a **sample** so that it is representative of the **population**. For example, if the population to be surveyed contains twice as many women as men then the sample should also contain twice as many women as men.

R

radian A unit of **angle** measure. The radian is the angle **subtended** at the centre of a **circle** by a **minor arc** of length equal to the **radius** of the circle. Therefore, one **degree** equals $\pi/180$ radians.

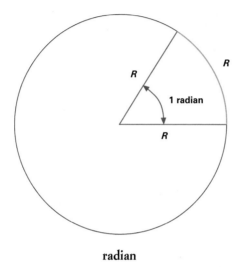

radian

radius (*pl.* **radii**) The distance from the centre of a **circle** to any point on its **circumference**. It is also used as a name for the straight line joining the centre of a circle to any point on the circumference.

random **1**. *Random sample*. A sample taken from a **population** is said to be random if every member of the population has an equal **probability** of being chosen.
2. *Random number*. A number in which each **digit** is chosen from a set of values 0, 1, 2, 3, 4, 5, 6, 7, 8, 9, where each value has the same chance of being chosen.

range **1**. The difference between the largest and smallest data values in a given set of data. For example, the range of 3, 2, 6, 2, 2, 4, 7, is the difference between 7 and 2; that is $7 - 2 = 5$. See also **interquartile range**.
2. The **set** of **images** produced when a given function acts on its **domain**. For example, if the domain of the function $y = x^2$ is restricted to the set $\{1, 2, 3, 4\}$ then the range is just $\{1, 4, 9, 16\}$.

rank A number assigned to a **data** value to express the position it appears at when the data is arranged in ascending **order**. For example, the data set {2, 9, 7, 7, 3, 4, 6} could be ranked by first putting them in ascending order:

$$2, 3, 4, 6, 7, 7, 9,$$

which means they are ranked as follows:

$$2(1), 3(2), 4(3), 6(4), 7(5.5), 9(7),$$

where the numbers in the brackets are the rank. Notice the rank of 7, which appeared twice, is joint 5th, which means that the value 9 becomes 7th in ranking.

ratio The ratio of A to B is just A divided by B and is written as $A{:}B$.
    For example, the ratio of £2.00 to 50p is written as £2.00:£0.50 = 4:1. Notice that the two quantities must both be in the same units if they are to be compared in this manner.

rational number A number that can be expressed as a **ratio** of two **integers**. For example, 2.5 is a rational number because it may be expressed as 5/2. Rational numbers can therefore be expressed as either **terminating** or **recurring decimals**. The **symbol** for the **set** of rational numbers is Q. See also **irrational number**.

rationalize Expressions containing nth-**roots** of a **variable** may be simplified by rationalizing to remove the root. For example, the expression:

$$x = \surd(2x - 1)$$

may be written as:

$$x^2 - 2x + 1 = 0.$$

In this example, the value of the expression has not changed but it has been rationalized to an expression that does not contain the **square root**.
    For an expression in which the root appears in the **denominator** of a **fraction**, an equivalent expression could be found by multiplying the top and bottom of the fraction by a **factor** to leave the root in the **numerator**. For example,

$$\frac{1}{y + \sqrt{x}} = \frac{1}{y + \sqrt{x}} \times \frac{\left(y - \sqrt{x}\right)}{\left(y - \sqrt{x}\right)} = \frac{\left(y - \sqrt{x}\right)}{\left(y^2 - x\right)}$$

shows how an expression may be rationalized to leave the root in the numerator.

ray A straight line extending from a **point**, called the **origin** of the ray. Rays are also sometimes referred to as *half-lines*, owing to the fact that the origin effectively **bisects** the **real number** line **parallel** to the ray.

real numbers The set of numbers that is the **union** of the **rational** and **irrational numbers**.

To each infinitesimal point on the continuous real number line there corresponds a real number. Therefore between any two real numbers the number of real numbers is **infinite**.

reciprocal The reciprocal of x is the number of times x divides into 1; that is, the reciprocal of x is $1/x$. For example, the reciprocal of 5 is $1/5 = 0.2$. The **product** of a number and its reciprocal is always 1.

rectangle A **quadrilateral** in which all the interior **angles** are equal to 90°. The opposite sides are of equal length. A rectangle has two lines of symmetry and a rotational **symmetry** of **order** 2. The **diagonals** of a rectangle **bisect** each other and also bisect the rectangle itself.

rectangle number A number having more than two **factors**. These numbers may be represented as rectangular configurations with the factors making up the dimensions of these rectangles. For example, the number 12 may be represented as a 3 by 4 rectangle or a 2 by 6 rectangle.

The set of all rectangle numbers is: {4, 6, 8, 9, 10, 12, 14, 15, 16, ...}. The **union** of the rectangle numbers with the **prime numbers** is the set of all **natural numbers**, apart from 1, which is unique in having just one factor.

rectangular 1. Describing a solid in which the cross section is a rectangle. A rectangular **prism** (also known as a cuboid) is a prism whose cross-section is a **rectangle**.
2. At right angles. Thus, the **axes** of a **Cartesian coordinate** system are often known as rectangular axes.
3. A *rectangular hyperbola* is a hyperbola whose **asymptotes** are the x- and y-axes.

The general **equation** of a rectangular hyperbola is $xy = c^2$; that is, the product of corresponding x- and y-coordinate values is always equal to a **constant**. See illustration on the next page.

rectilinear motion Motion in a straight line. For example, light in a uniform medium is sometimes said to propagate rectilinearly; that is, travel in straight lines.

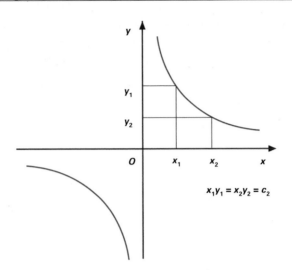

$$x_1y_1 = x_2y_2 = c_2$$

rectangular: *a rectangular hyperbola*

recurring decimal The **decimal** representation of a **rational number** may contain an infinitely repeating sequence of decimal **digits**. Such a number is called a recurring decimal. **Fractions** containing **denominators** containing **prime factors** other than 2 and 5 will have recurring decimal representations. For example:

$$1/3 = 0.333333333...,$$
$$1/6 = 0.166666666...,$$

See also **terminating decimal**.

re-entrant polygon A re-entrant **polygon** is one in which one or more of the **interior angles** is greater than 180°; i.e. one or more interior angles is a **reflex** angle.

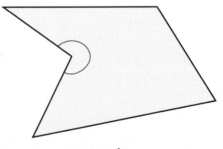

re-entrant polygon

reflection An **operation** that conserves **angles**, lengths and **areas** but reverses the object with respect to a coordinate **axis**. In two dimensions the vertices of a figure may be represented as 2 × 1 column **matrices**. In this representation a reflection of the figure through a *mirror line* would be represented by a 2 × 2 matrix. Below, the matrix corresponds to a reflection in the *x*-axis:

$$\begin{pmatrix} 1 & 0 \\ 0 & -1 \end{pmatrix}$$

In the diagram, the point $x = 1$, $y = 1$, is reflected in the *x*-axis to form the point $x = 1$, $y = -1$.

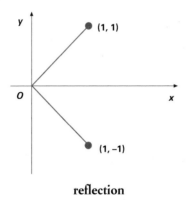

reflection

reflex angle An **angle** that is greater than 180° and less than 360°.

reflexive relation A **relation** on a **set** where every **member** of the set is related to itself.

region 1. The areas of a **plane** that have been defined by the **arcs** of a **network**.
2. Regions of the *x–y* plane can be represented by **inequalities**. For example, the shaded region in the diagram overleaf may be represented by the following inequality:

$$x \geq 2 - 2x$$

regular Describing **polygons** that have sides of equal size or **polyhedra** that have faces of equal shape.

For example, the **Platonic solids** are regular polyhedra made of regular polygons: The **cube** is a regular polyhedron with six **squares** as faces.

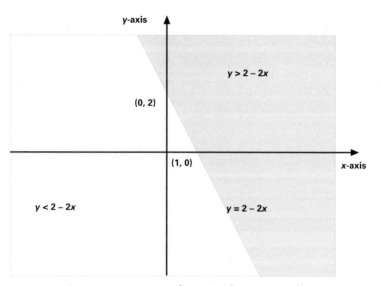

region: *representation of a region by an inequality*

relation A connection between **members** of **sets**. **Functions** may be thought of as relations between a set called the **range** and a set called the **image** of the range.

remainder When a **dividend** a is divided by a **divisor** b the result is in general a **quotient** plus a remainder. For example, 17 divided by 5 is the quotient 3 with a remainder of 2.

remainder theorem The theorem that if polynomial $P(x)$ is divided by $(x - a)$ the remainder is $P(a)$. For example, if

$$P(x) = x^2 + 2x + 1,$$

and we divide $P(x)$ by $(x - 2)$, then the remainder of the division is:

$$P(2) = 4 + 4 + 1 = 9.$$

Notice that if we divide $P(x)$ by $(x + 1)$, then P(-1) = 0; that is $(x + 1)$ is a **factor** of $P(x)$ and therefore $P(x)/(x + 1)$ has no remainder.

repeated root When a **polynomial** is **factorized**, repeated factors mean that there are repeated roots. For example, the polynomial:

$$x^3 - 3x + 2 = 0$$

may be factorized as:

$$(x - 1)(x - 1)(x + 2) = 0,$$

which means that the roots $x = 1$ are repeated roots. This corresponds to the point at which the curve $y = x^3 - 3x + 2$ just glances the x-axis at $x = +1$.

resolve **1.** To find the **components** of a **vector** with respect to a particular **coordinate** system.
2. To make an observation that leads to a measurement. See **accuracy**.

resultant A vector equivalent to the combined effect of a number of **vectors**. A vector itself may be considered to be the resultant of the vector **addition** of its **components**.

rhombus A **parallelogram** that has all its sides equal. A rhombus has two lines of **symmetry** and a rotational symmetry of **order** 2. The diagonals of a rhombus **bisect** each other at **right angles** through the **centre of symmetry** of the figure.

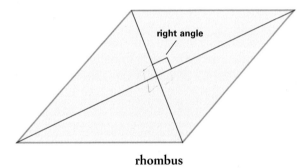

right angle

rhombus

right angle An **angle** equal to 90 **degrees**, or $\pi/2$ **radians**. In a diagram, a right angle between two intersecting lines is denoted as such by a small square. See diagram above.

root The roots of an **equation** in an unknown are the values of the unknown which satisfy the equation. For example, the following equation in x:

$$x^3 - 3x + 2 = 0,$$

is true for $x = 1$, 1, and -2. These values of x are called the roots of the equation $x3 - 3x + 2 = 0$.

rotation A geometrical **transformation** in which every point on a figure is rotated through an **angle** with respect to a **coordinate system**.
 Such a transformation may be represented as the action of a **matrix** on the coordinates of the figure. For example, a figure could be rotated

through 90° about the origin of a **Cartesian coordinate** system by the following matrix transformation:

$$\begin{pmatrix} x' \\ y' \end{pmatrix} = \begin{pmatrix} 0 & -1 \\ 1 & 0 \end{pmatrix} \begin{pmatrix} x \\ y \end{pmatrix}$$

where (x, y) and $(x¢, y¢)$ are respectively the coordinates of the original and rotated figures.

rounding An **approximation** for a number that is accurate enough for some specific purpose. The difference between the number before and after rounding is called the *rounding error*. For example, the number 1.125 may be rounded to 1.13 with a rounding error of 1.13 − 1.125 = 0.005.

The effects of rounding errors can be seen in calculations involving **recurring decimals** or **irrational numbers**. For example,

$$(2 \div 3) \times 3 \approx 0.67 \times 3 = 2.01 \neq 2$$

that is, 2 divided by 3 is actually a recurring decimal but can be represented by the rounded figure of 0.67. In taking this value, one has actually introduced the rounding error into the result so multiplying back by 3 does not result in 2 but 2.01.

row A list of numbers or letters written horizontally. For example 1 2 3 4 5. The **order** of an arbitrary **matrix** is given as $n \times m$, where n denotes the number of rows in the matrix.

S

sample The part of a **population** that one is considering within a statistical analysis. The act of taking a sample within a **population** is called *sampling*. There are two factors that need to be considered when sampling from a population:
(1) The size of the sample. The sample must be large enough compared to the population for the results of a statistical analysis to have any significance.
(2) The manner in which the sampling is done. The sample should be representative of the population. Any variations of characteristics across the population should be accounted for in the sample. For an example of this see **quota sampling**. Sometimes there may be no information about the characteristics of a population. In such cases a sample is taken in which all items are equally likely to be chosen. This is a *random* or *cluster sample*.

scalar A quantity that has **magnitude** but no direction. Examples of quantities that are scalars are mass, temperature intervals, and energy.
It is sometimes useful to think of a scalar as a 1×1 **matrix**. In this way one can easily visualize the **multiplication** of matrices by scalars. For example, the column matrix that represents an arbitrary **vector** in the x–y-plane may be multiplied by a scalar, k, as follows:

$$k \begin{pmatrix} x \\ y \end{pmatrix} = \begin{pmatrix} kx \\ ky \end{pmatrix}$$

This effectively multiplies the length of the vector by the factor k.

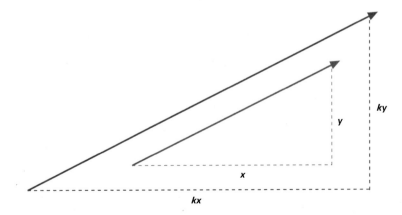

scalar: s*calar multiplication of a vector by k*

scale factor A factor by which the physical dimensions of a figure are increased in an **enlargement** of a figure.

The effect of a negative scale factor is not so obvious. When one applies a **negative** scale factor the **image** of a point will be found to be on the other side of the **centre of enlargement**. For example, in the diagram the line AB is enlarged by a factor of –2 to the image line $A'B'$.

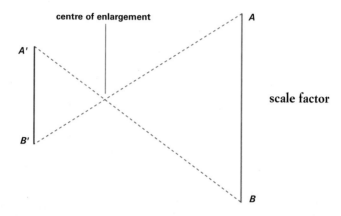

scalene Describing a **triangle** in which no two sides or **angles** are **equal**.

scatter graph If a set of data has two or more aspects to it, then a scatter graph may be used to display any relationships which may exist between any two aspects. For example, both the height and the weight of children in a school may be displayed on a scatter graph. One could then see with relative ease whether or not there is a **correlation** between these two aspects. See also **line of best fit**.

scientific notation See **standard form**.

search method A method of solving for the **roots** of numbers by making successively better **estimates**.

For example, in the **equation** $x^2 = 5$ an initial guess at an answer for x might be as crude as saying that it must be between 2 and 3 (since $2^2 = 4$ and $3^2 = 9$). Following the search method, the next estimate to try is the **mean value** between 2 and 3; that is 2.5. This is found to be too large and becomes the upper value for the next estimate between 2 and 2.5, i.e. a value of 2.25. Squaring this gives 5.0625, which is too high. The next value to be tried is the mean of 2.25 and 2.125, i.e. 2.1875. Squaring this gives 4.7852, which is too low. Repeated use of this method **iterates** towards a solution for x; by the fifth iteration the value is approaching the square root of 5.

second 1. A unit of time. There are 60 seconds in one **minute**.
2. A measure of **arc** or **angle**. There are 60 seconds of **arc** in 1 **minute** of arc; that is, $60'' = 1'$.

sector A **region** of a **circle** bounded by an **arc** and two **radii**. The **area** bounded within a sector is easily calculated by considering the region as a part of the whole circle which has an area of πR^2, where R is the radius. The area of the sector is therefore:

$$A = (\theta/360) \times \pi R^2, \text{ or}$$
$$A = (\theta/2\pi) \times \pi R^2,$$

where the first and second expressions respectively express the angle θ in **degrees** and **radians**.

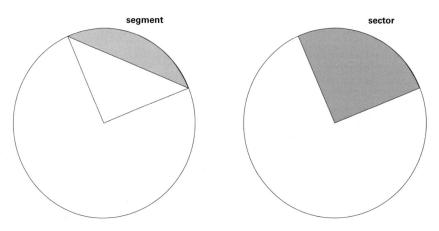

segment and sector

segment A **region** of a **circle** bounded by an **arc** and a **chord**. The area within the segment of a circle is easily calculated by first calculating the area of the surrounding **sector** and subtracting from this value the area of the enclosed **triangle**.

semi-interquartile range Half of the **interquartile range**.

sequence A **progression** of **values** which are related either by a **common difference** or **ratio**.
1. *Arithmetic sequence.* A sequence that is generated by the **addition** or **subtraction** of the same number each time. For example, the general term:

$$u(n) = 3n + 1$$

describes the following sequence:

$$4, 7, 10, 13, 16, \ldots,$$

for $n = 1, 2, 3, 4, \ldots,$ etc.

2. *Geometric sequence*. A sequence that is generated by the multiplication or division of the same number each time. For example, the general term:

$$u(n) = 4^n$$

describes the following sequence:

$$4, 16, 64, 256, 1024, \ldots,$$

for $n = 1, 2, 3, 4, \ldots,$ etc.

series A series is a **sum** of terms of a **sequence**. For example, associated with the sequence 1, ½, ¼, …etc, there is a series $1 + ½ + ¼ + \ldots$etc. One may consider either a sum to n terms of a series or a sum to **infinity**.

set A collection of distinct things. These things are referred to as **members** or **elements**, which can be finite or **infinite** in number. The **symbol** used to denote a set is {} and the symbol $\in$ means 'is a member of'.

shear An abstract **transformation** of a plane figure in which one line remains fixed and all other points are displaced parallel to this fixed line. The distance a point is displaced by is found by multiplying the factor of the shear by the distance of the point from the fixed line. A shear does not leave **angles** invariant nor does it preserve some lengths.

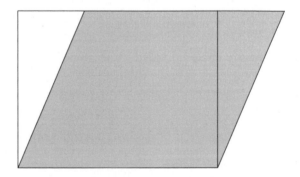

shear: *transformation of a square into a parallelogram*

side One of the line **segments** forming the boundary of a **polygon**. For example, a **quadrilateral** has four sides.

sigma (Σ) The Greek letter for capital sigma is an important mathematical symbol used to denote 'the **sum** of'. For example, the sum of the first five terms of the **sequence** whose general term is

$$u(n) = 3n + 1 \text{ is:}$$

$$4 + 7 + 10 + 13 + 16 = 50,$$

which is most conveniently represented by the expression:

$$\sum_{n=1}^{n=5} (3n + 1) = 50$$

This incorporates the sigma symbol with superscripts and subscripts indicating the values of n to be used in the sum.

sign A notation used as part of the description of a quantity (e.g. +, –, $\sqrt{}$) or to indicate an **operation** to be performed (e.g. +, ×, Σ). See also **directed**.

significant **1**. In statistics, observations are said to be significant if they vary enough from hypothesized values that they are unlikely to be the result of random fluctuations.
2. The significance of a particular digit in a number is concerned with its relative size in the number. For example, in the number 78.09 the digit 7 is most significant to its value (corresponds to 7' 10) and 9 is the least significant (corresponds to 9' 1/100). See also **approximation**.

similar Describing two bodies that share the same interior angles. Thus two bodies that have the same shape but differ in size are still said to be similar.

When a shape is enlarged the shape and its image are similar. The corresponding angles in similar figures are equal and the ratios of the lengths of corresponding sides is constant.

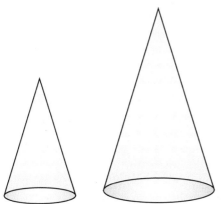

similar: *two similar cones*

simple interest See **interest**.

simultaneous equations **Equations** of more than one unknown which all must be 'simultaneously' satisfied. To solve a set of simultaneous equations there must be as many equations as unknowns. There are many methods for solving sets of simultan- eous equations. A few are listed below using a simple example of two **linear** equations (A) and (B):

$$3x - 4y = 23 \text{ (A)}$$
$$4x + 3y = 14 \text{ (B)}$$

(1) *Elimination method.* This depends on eliminating one of the unknowns. For example, the simultaneous equations (A) and (B), may be modified without changing their value by multiplying (A) by 3 and (B) by 4 to leave:

$$9x - 12y = 69 \text{ (A}')$$
$$16x + 12y = 56 \text{ (B}')$$

Now both equations contain $12y$, so if the equations are added together one is left with $25x = 125$, which means that $x = 5$. Substitution of this value into either of the two original equations leads to $y = -2$.

(2) *Substitution method.* This involves the solving of one of the unknowns in terms of the others in one equation and substituting this into another equation. For example, in equation (A), x may be rearranged in terms of y as follows:

$$3x - 4y = 23 \rightarrow x = (23 + 4y)/3.$$

This may then be substituted into equation (B) to give an equation for y, which may be substituted into the equation for x to obtain $x = 5$.

(3) *Graphical method*: This involves the plotting of the **graphs** of the simultaneous equations (see **Appendix 1**). The **coordinates** of the **intersections** of these graphs are the solutions to the unknowns, which satisfy the equations simultaneously.

(4) *Solution by matrices*: This involves the representation of the equations (A) and (B) as one **matrix** equation with a **coefficient** matrix (C):

$$(C)\begin{pmatrix} x \\ y \end{pmatrix} = \begin{pmatrix} 3 & -4 \\ 4 & 3 \end{pmatrix}\begin{pmatrix} x \\ y \end{pmatrix} = \begin{pmatrix} 23 \\ 14 \end{pmatrix}$$

By matrix multiplication this matrix equation leads to the two linear equations (A) and (B). Pre-multiplying both sides of this matrix equation by the **inverse** of (C) leads to the matrix equation:

$$\left(C^{-1}\right)(C)\binom{x}{y} = \frac{1}{25}\begin{pmatrix} 3 & 4 \\ -4 & 3 \end{pmatrix}\begin{pmatrix} 3 & -4 \\ 4 & 3 \end{pmatrix}\binom{x}{y}$$

$$= \begin{pmatrix} 1 & 0 \\ 0 & 1 \end{pmatrix}\binom{x}{y}$$

$$= \frac{1}{25}\begin{pmatrix} 3 & 4 \\ -4 & 3 \end{pmatrix}\binom{23}{14}$$

$$\rightarrow \binom{x}{y} = \binom{5}{-2} \rightarrow x = 5, y = -2$$

sine A **function** used in **trigonometry**. In a **right-angled triangle** the sine of an **angle** is calculated by the **ratio**:

$$\sin (\theta) = \text{opposite} \div \text{hypotenuse},$$

where *opposite* is the side opposite the angle θ. The graph of the sine function is given in **Appendix 1**.

sine rule For a **triangle** *ABC*, the sine rule is given by the following expressions:

$$a/\sin (A) = b/\sin (B) = c/\sin (C).$$

See illustration.

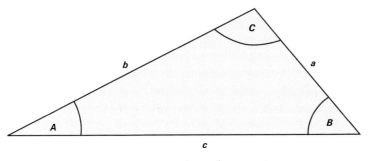

sine rule

skew Describing a **histogram** that is not **symmetric** about the **mean** value.
   In the diagram, the two histograms labelled A and B both have 'tails'. A has a tail to the left whilst B has a tail to the right. Distribution A is said to be *skewed to the left* and distribution B is said to be *skewed to the right*. A distribution that is skewed to the right is said to be *positively skewed* whilst a distribution that is skewed to the left is said to be *negatively skewed*.
   As an example of skewed distributions, consider a test which was found to be easy by one group of pupils and hard by another group. The mark

distributions would be negatively skewed if the group found it easy and positively skewed if the group found it hard. Skewed distributions are by definition not symmetric about the mean therefore the mean, **mode**, and **median** do not coincide.

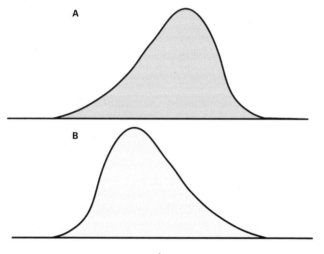

skew

small circle A circle drawn on the surface of a **sphere**, whose **centre of symmetry** does not coincide with the centre of the sphere. Small circles whose centres lie on the Earth's axis are all lines of **latitude**.

solids The relationship between **similar** solids or figures is usually given in terms of a ratio of corresponding edge lengths (or linear **dimensions**). The ratio of corresponding total surface areas or total volumes can be considerably more difficult to visualize.

Area is measured in **square** units. Therefore if the ratio of the edge lengths is $A : B$ then the ratio of the total surface areas is $A^2 : B^2$.

Volume is measured in **cubic** units. Therefore if the ratio of the edge lengths is $A : B$ then the ratio of the total volumes is $A^3 : B^3$.

solution The result of solving a mathematical problem. Solutions are often given in **equation** form.

speed The total distance travelled (d) by a particle divided by the time taken to complete the journey (t) is called the *average speed* (V); that is, $V = d/t$.

The **gradient** of a **distance–time graph** at a particular time is the *instantaneous speed* at that instant.

sphere A **solid** with full three-dimensional rotational **symmetry**. This means that all points on the surface of a sphere are a fixed distance from the **centre of symmetry** of the sphere. This fixed distance is called the **radius** of the sphere. The surface **area** (S) and **volume** (V) of the sphere may be expressed in terms of the radius (R) in the following formulae:

$$S = 4\pi R^2,$$
$$V = 4\pi R^3/3.$$

square **1.** The result of multiplying a number by itself once. For example, 2^2 or 2 squared is equal to $2 \times 2 = 4$.
2. A **rectangle** with equal sides.

square numbers Numbers whose square roots are $+n$ or $-n$ where n is a **natural number**. For example, 1, 4, 9, 16, 25, ..., etc are square numbers.

square root A number whose square is a given number.
Taking the *square root* of a number a is the process of finding the numbers $+b$ and $-b$ that when multiplied by themselves give a. For example, the square roots of 4 are +2 and –2; that is $(+2) \times (+2) = 4$ and $(-2) \times (-2) = 4$. Taking the square root of a negative number leads to multiples of **imaginary** numbers. See also **complex number**.

standard deviation See **deviation**.

standard form A convenient way of writing numbers that are very large or very small. In this notation, any number is written as a multiple of **powers** of the number 10.
    For example, the number 375 000 000 (three hundred and seventy five million) may be expressed more conveniently as:

$$3.75 \times 10^8.$$

Similarly, very small numbers may be more conveniently expressed in standard form. For example, 0.000 375 is:

$$3.75 \times 10^{-4}.$$

stationary point A point on the **graph** of a **function** at which the **gradient** of the function is **zero**; i.e., the **tangent** to the function at this point is **parallel** to the *x*-axis. See also **Appendix 1**.

statistics The study of methods for analysing large quantities of **data** that has been **sampled** from a **population**.

straight line The **locus** of a point that follows a straight line may be represented by the general **equation**

$$y = mx + c,$$

where (x,y) are the **coordinates of the point**, m is the **gradient** of the line, and c is the **intercept** on the y-axis. See **Appendix 1**.

stretch An abstract **transformation** of a plane figure in which lengths and **areas** are varied but **angles** are conserved. To define a stretch one need only specify the factor or amount of stretch and the direction of the stretch. Stretches are usually **parallel** to one of the **coordinate** axes. For example, the **circle** in the diagram has been stretched parallel to the x-axis to form the **ellipse**.

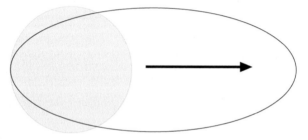

stretch: *transformation af a circle into an ellipse*

subject of a formula The subject of a **formula** is the quantity that the formula is equated to. For example, the formula for the time period T of a simple pendulum of length l is:

$$T = 2\pi\sqrt{(l/g)}.$$

where g is the acceleration due to gravity ($g = 9.81$ m/s2). In the formula T is the subject. The formula may be rearranged according to the rules of **algebra** to a form in which l is the subject:

$$T^2 = 4\pi^2 l/g \;\rightarrow\; l = gT^2/4\pi^2.$$

See **Appendix 2**.

subset A **set** within a set. For example, the set {2, 3, 4} is a subset of the set {1, 2, 3, 4, 5, 6}.

subtend The angle subtended at a point by a line or arc is the angle between lines running radially from the point to the extremities of the line or arc.

subtraction 1. The **operation** in **arithmetic** that determines the **difference** of two **scalars**.
2. The difference of two **vectors** $a - b$ can be thought of as the vector addition of the vector a to the vector $-b$.

Subtraction is the **inverse** operation to **addition**.

sum The result of the **addition** of two or more quantities.

surd An irrational **root** of a **rational number**. For example:

$$\sqrt{3}, \ \sqrt{11}, \ 3\sqrt{4},$$

are all surds. For the properties of surds under algebraic operations, see **Appendix 2**.

surface area The **area** of the surface that bounds a **solid**. The surface areas of some familiar solids are given in the table below:

Solid	*Surface area*
Cylinder (radius r, height h)	
area of curved surface	$2\pi rh$
area of two ends	$2(\pi r^2)$
area including both ends	$2\pi rh + 2\pi r^2$
Cone (radius r, slant height l)	
area of curved surface	πrl
area of base	πr^2
area including base	$\pi rl + \pi r^2$
Sphere (radius r)	$4\pi r^2$
Hemisphere (radius r)	
area of curved surface	$2\pi r^2$
area of base	πr^2
area including base	$2\pi r^2 + \pi r^2 = 3\pi r^2$

survey The act of collecting **data** for a statistical analysis. When conducting a survey, the following points may help:
(1) Ask clear and concise questions.
(2) Decide how your data is to be collated and analysed.
(3) Allow for all possible answers to your questions.
(4) If your survey is asking for opinions, make sure that your opinion is not evident in the way you word your questions
(5) Keep your questionnaire short.
See **Appendix 3**.

symbol A sign to denote a quantity, operation or relation. For example: $\times, \div, +, -$.

symmetry A figure is said 'to have symmetry' under a certain transformation if the aspect of the figure remains unchanged under that transformation.

Line/plane of symmetry. A line/plane that divides a figure/solid into two **congruent** figures/solids.

Reflective symmetry. A figure/solid has reflective symmetry if it has a line/plane of symmetry.

Rotational symmetry. A figure/solid has rotational symmetry if it coincides with itself more than once under a 360 **degree** rotation about a **centre of symmetry/axis**. The number of times the shape coincides with itself under a complete rotation is called the **order** of the rotational symmetry.

symmetric relation A relation * is said to be symmetric on a set S if: $\forall a,b \in S$, a*b then b*a.

T

tangent **1**. A **trigonometric** function. In a **right-angled** triangle the tangent of an **angle** is calculated by the **ratio**:

$$\tan(\theta) = \text{opposite} \div \text{adjacent},$$

where *opposite* and *adjacent* are respectively the sides opposite and adjacent to the angle θ. The graph of the tangent function is given in **Appendix 1**.
2. A straight line that just touches a given **curve** at a particular point. See also **gradient**.

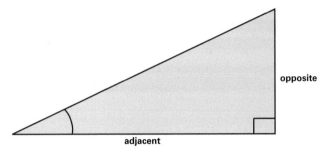

tangent

terminating decimal The **decimal** resulting from the **division**:

$$5 \div 8 = 0.625$$

is said to be a terminating decimal since it terminates (stops) after three decimal places.

tessellation A shape is said to be *tessellated* if, when it is **translated** and/or **reflected** and/or **rotated**, it completely fills a space leaving no gaps. A space filled in this manner is said to form a *tessellation*.

tetrahedron A **Platonic solid**. A tetrahedron is a **polyhedron** constructed from four **equilateral** triangular faces.

theorem A proposition to be proved which is constructed from **hypotheses** and **axioms**. See also **Appendix 3**.

tonne A unit of mass equivalent to 1000 **kilograms**. See **unit conversions**.

trajectory The path of a moving body. For example, if the body is horizontally projected in the Earth's gravitational field the ensuing path takes the form of a **parabola**.

transcendental number An **irrational number** that cannot be obtained as a result of solving a **polynomial** with **rational coefficients**. For example π and *e* are transcendental numbers.

Numbers that can be obtained through algebraic means are said to be *algebraic*; that is, solutions of equations such as: $x^2 = 2$ leading to $x = 1.414$, etc., are *algebraic numbers*.

transformation The following are examples of transformations:

Translation: translation moves an object from one place to another while maintaining orientation, size and shape. A translation may be achieved on a plane figure if the coordinates of each **vertex** is shifted by the same **displacement vector**.

Enlargement: enlargement increases or de-creases size but maintains shape. See also **scale factor**.

Rotation: rotation conserves **angles** but alters the orientation of the figure. To define a rotation one requires three pieces of information: (1) the position of the centre of rotation; (2) the direction of rotation; and (3) the angle through which the figure is to be rotated.

Reflection: reflection conserves angles, lengths and therefore **areas** but reverses the figure with respect to a **coordinate axis**.

Stretch: transformation that conserves angles but changes lengths and therefore areas.

Shear: transformation in which a line or a **plane** remains fixed whilst other points move, **parallel** to the fixed line or plane.

transitive relation A relation * is said to be transitive on a set S if:

$$\forall a,b \in S$$
$$a*b, b*a \Rightarrow a*c$$

translation A transformation in which all **points** of a plane figure are displaced by the same **displacement vector**. In the illustration opposite, the plane figure ABCDEF is translated to A′B′C′D′E′F′.

transversal A line that **intersects** two or more other lines.

trapezium (pl. **trapezia**) A **quadrilateral** with only one pair of **parallel** sides. See also **area**.

trapezium rule See **area under a curve**.

tree diagram A diagrammatic tool used in **probability** work. Events are represented as points and are related to each other by a branching of lines.

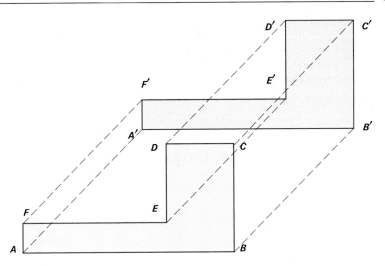

translation

trial The repetition of a particular experiment under controlled conditions in a statistical investigation. See **statistics**.

trial and improvement methods See **search method**.

triangle A three-sided **polygon**. Triangles may be classified as:
(1) **scalene:** no two sides of these triangles are equal in length.
(2) **isosceles:** two of the sides of the triangle are equal in length.
(3) **equilateral:** all the sides of these triangles are equal in length.
A **right-angled** triangle has an interior **angle** equal to 90 degrees.

trigonometry A branch of mathematics concerned with the study of the properties of the *trigonometric* functions **sine, cosine,** and **tangent,** and their **reciprocals.** See **Appendix 1.**

trinomial Any **polynomial** with just three terms. For instance:

$$x^5 + 3x^2 + 5$$

is an example of a trinomial.

turning point A **point** on the **graph** of a **function** which is a local **maximum** or **minimum** value of the function. The **gradient** of a **tangent** at a turning point is zero; that is, tangents taken at these points are **parallel** to the *x*-axis. See also **stationary** points and **Appendix 1.**

U

unbounded function A **function** is said to be unbounded if over an **interval** the **value** of the function can be found to exceed any value no matter how big; that is, the function tends towards an **asymptote** at **infinity** at a particular value of the **variables**. For example, the function in the diagram is $y = 1/x$. Over the interval x = [0, 1] the function is seen to be unbounded; that is, for x equal to a small and positive number, y is equal to a large and positive number (see **Appendix 1**).

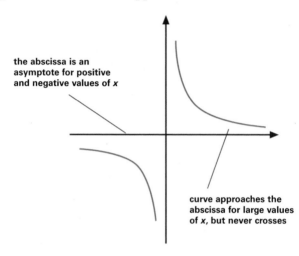

the abscissa is an
asymptote for positive
and negative values of **x**

curve approaches the
abscissa for large values
of **x**, but never crosses

unbounded function

unconditional inequality An **inequality** that is true for all values of the **variable**. For example:

$$x + 3 > x$$

is an unconditional inequality, true for all x.

union An **operation** on **sets**. The union of two sets A and B is denoted $A \cup B$ and is the set of all **elements** contained in A together with the elements contained in B. For example, if $A = \{1, 2, 3, 4\}$ and $B = \{3, 4, 5, 6\}$ then the union is:

$$A \cup B = \{1, 2, 3, 4, 5, 6\}.$$

unit **1**. Denoting unity; that is the **value** 1. For example a 'a circle of unit radius' is a **circle** of **radius** 1 of whatever unit is being employed.
2. A standard measure of a quantity e.g. mass, length, time (see **dimension**). For example, the **kilogram** is a unit of mass (see **unit conversions**).

unit conversions There are two main systems of **units** which are generally used; these are the **Imperial** and **metric** systems. The list below shows unit conversion factors for the quantities length, area, volume and mass/weight.

Length:
1 m = 100 cm = 1000 mm = 10^6 μm = 10^9 nm
1 km = 1000 m = 0.6214 miles
1 m = 3.281 ft = 39.37 in
1 in = 2.540 cm
1 ft = 30.48 cm
1 yd = 91.44 cm
1 mile = 5280 ft = 1.609 km
1 nautical mile = 6080 ft
(Key: m metre, cm centimetre, mm millimetre, μm micrometre, nm nanometre, km kilometre, ft feet, in inches, yd yard).

Area:
1 cm^2 = 0.155 in^2
1 m^2 = 10^4 cm^2 = 10.76 ft^2
1 in^2 = 6.452 cm^2
1 ft^2 = 144 in^2 = 0.0929 m^2

Volume:
1 litre = 1000 cm^3 = 10^{-3} m^3 = 0.03531 ft^3 = 61.02 in^3
1 ft^3 = 0.02832 m^3 = 28.32 litres = 7.477 gallons
1 gallon = 3.788 litres

Mass/Weight:
1 kg has the weight of 2.205 lb
1 lb = 16 oz
14 lb = 1 stone
   therefore
1 stone = 6.349 kg
(Key: lb pound, oz ounce).

universal set The **set** containing all **elements** of all the sets that are being considered in a mathematical problem. All other sets are **subsets** of the universal set. It is represented by the symbol $\mathscr{E}$.

upper bound The upper bound of a **set** of numbers is the number that no other **members** of the set exceed. For example, the function in the diagram has an upper bound of $f(x) = 1$.

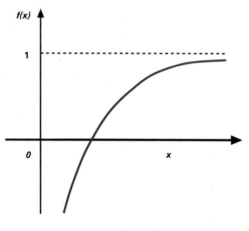

upper bound

upper quartile See **quartile.**

V

value Amount or quantity denoted by algebraic term or expression. The evaluation of a **function** of a **variable** at a particular value of the variable is achieved by introducing the value of the variable into the **equation**. For example:

$$f(x) = 3x + 4$$

may be evaluated for $x = 2$ giving:

$$f(2) = 3 \times (2) + 4 = 10;$$

that is, $f(2) = 10$.

See also **absolute value**.

variable A quantity that can assume a **range** of **values**. These values may be **discrete** or **continuous** in nature. Variables which may take discrete values only are called *discrete variable*. Variables which may take any value within a given range are called *continuous variables*.

A function of a variable is itself in a sense a variable; that is, if $y = f(x)$ then the variable y is a function of the variable x. For example, consider a physical quantity Q which is a function of a variable T according to the equation:

$$Q = m \times C \times T,$$

where the quantities m and C are **constants**. There are three kinds of variable in this equation:

(1) The *dependent variable*: Here, Q is referred to as the dependent variable since in its present form the equation considers changes in Q dependent on the value of T. The values of Q are said to belong to the **range** of the function $Q\,(T)$.

(2) The *independent variable*: T is referred to as the independent variable because it is changes in T which correspond to a change in Q. The values of T are said to come from the **domain** of the function $Q(T)$.

(3) The *control variables*: Although the quantities m and C are constants in the equation, they are nevertheless often considered as control variables. If the equation were to represent an equation relating variables in a physics experiment, the control variables are the variables which will insure a *fair test* is carried out.

See also **Appendix 3**.

variance For n items of **data** x_i with a **mean** of $<x>$, the mean of the squares of the deviations of each x_i from the mean $<x>$, is called the variance σ^2; that is:

$$\sigma^2 = \frac{\sum_{i=1}^{n}\left(x_i - <x>\right)^2}{n}$$

The **square root** of the variance is known as the standard **deviation**.

vector A quantity that requires both a **magnitude** and a direction for a complete description. Two examples of vectors are **displacement** and **velocity**. Vector quantities in text books are often represented in bold type e.g. a vector *a*.

Vectors in a **plane** or in three **dimensions** may be represented as **column matrices**. For example, the displacement of a knight on a chess board is illustrated below as a vector in the plane of the board:

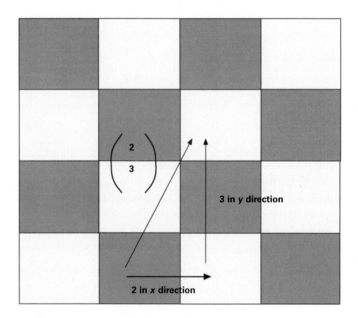

vector: *a column vector*

As column matrices, the **addition** of vectors follows the rules of the addition of matrices; that is, corresponding components are added together:

$$\begin{pmatrix}1\\2\end{pmatrix}+\begin{pmatrix}3\\4\end{pmatrix}=\begin{pmatrix}1+3\\2+4\end{pmatrix}=\begin{pmatrix}4\\6\end{pmatrix}$$

This leads to a simple rule for adding vectors. Since the effect of a combination of vectors is the effect of the combination of their components, the addition of two vectors A and B may be obtained by placing them 'head to tail' to form a **resultant**. For example, in the illustration the two vectors A and B are combined 'head to tail' to form a resultant R. See also **component** and

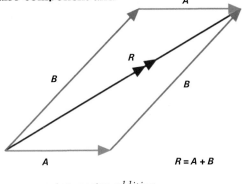

vector: *vector addition*

velocity Rate of change of **displacement** per unit time. Since displacement is a **vector** quantity, velocity is also a vector quantity.

Venn diagram A diagram showing the relationships between **sets**, representing them as **regions** enclosed by simple closed curves. The illustration shows examples of Venn diagrams for two sets A and B showing

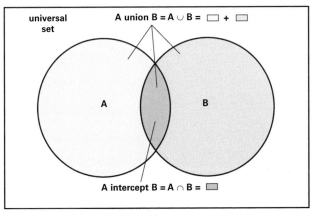

Venn diagram

the **universal set**, the **intersection** of A and B $(A \cap B)$, and the **union** of A and B $(A \cup B)$.

vertex (*pl.* **vertices**) Points at which sides meet on **polygons** and planes meet on **polyhedra**. The relationship between the number of vertices and the number of sides/planes on the polygons/polyhedra is **Euler's formula**. See also **Appendix 3**.

vertical The direction **perpendicular** to the **plane** of the **horizontal**.

volume The measure of the amount of space occupied by a **solid**. The volumes of some of the more familiar solids are given below:

Cone of radius R and height h: $(1/3) \times \pi R^2 \times h$
Cuboid of length l, height h and breadth b: $l \times h \times b$
Cylinder of radius R and height h: $\pi R^2 \times h$
Hemisphere of radius R: $(2/3) \times \pi \times R^3$
Pyramid of base area A and height h: $(1/3) \times A \times h$
Sphere of radius R: $(4/3) \times \pi R^3$
Prism of base area A and height h: $A \times h$

vulgar fraction A fraction that has integers for its **numerator** and **denominator**. Any **rational** number can be expressed as a vulgar fraction. For example, ½, ¼, ⅓, etc.

W

whole numbers **1.** Formally, the **positive** and **negative** counting numbers with **zero**:

$$\ldots -2, -1, 0, +1, +2, +3, +4, +5, \ldots.$$

2. In general usage, the positive counting numbers:

$$1, 2, 3, 4, 5, 6, \ldots..$$

Sometimes zero is also included in this sense as a whole number.

Y

yard An **Imperial unit** of length equal to three **feet**. There are 1760 yards in one **mile**.

Z

zero The **cardinal** number associated with the **empty set**. Zero is represented by the numeral 0 and is the number that obeys:

$$1 + 0 = 0 + 1 = 1.$$

zone A **region** of a **sphere** bounded by two **parallel** cutting **planes**. The surface area of a zone of a sphere of radius R, produced by cutting planes distance h apart is: $2\pi R h$.

Appendices

Appendix 1

Graph Sketching

It is often very useful to visualise the **graph** of a **function** without having to plot out the **points** accurately. For example, a quick sketch of the **boundaries** of an **inequality** can be of great help when solving for the **values** that satisfy the conditions. For an example of this, see the entry for **linear programming**.

There are a number of steps to be taken when sketching a function. The procedure presented here is just one way of ordering these steps in a methodical manner.

Step 1:
Determine the general shape of the function from the form of the **equation** which relates the **variables**. The diagrams following give a brief summary of forms and their corresponding equations. Note that the functions used in **trigonometry** are **periodic** and that the x-**axes** in these cases are graduated in **radians** ($\pi/2$ radians = $90°$).

Step 2:
Determine the **intercept** points on the x and y-axes.

Step 3:
Determine what happens to the function for x = large and positive, and for x = large and negative.

Step 4:
Join these points togther bearing in mind the general form of the function from the first step.

As an example of the application of these steps, the following analysis and sketch for a general **parabola** can be considered.

A parabola is the **curve** that results from plotting a **quadratic equation**. The general quadratic:

$$y = ax^2 + bx + c,$$

is a curve whose form depends on the constants a, b, and c. Neverless, such an equation has the same general form as the function

$$y = x^2,$$

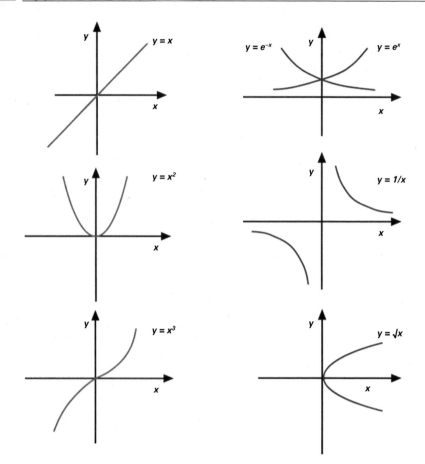

which has only one **turning point** or **stationary point**. The application of step 1 therefore leads one to conclude that the curve must have only one turning point and it must be **symmetric** about a vertical line through this turning point.

Application of step 2 may be achieved by using the **common formula** (assuming the common formula can be applied for the coefficients a, b, and c). The **roots** of the equation, that is, the points of intersection with the x-axis are given by:

$$x_0 = -b \pm \sqrt{(b^2 - 4ac)}/2a$$

where x_0 are the roots. The intersection of the curve with the y-axis is found by considering the equation for $x = 0$; this leads to a y intersection of c.

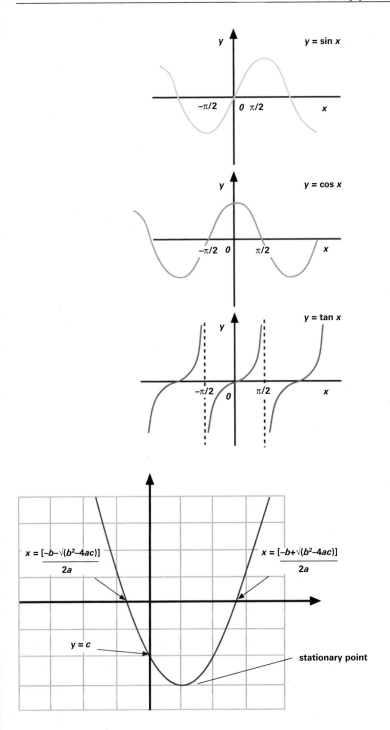

$y = \sin x$

$y = \cos x$

$y = \tan x$

$$x = \frac{[-b-\sqrt{(b^2-4ac)}]}{2a}$$

$$x = \frac{[-b+\sqrt{(b^2-4ac)}]}{2a}$$

parabola

$y = c$

stationary point

Calculus

Calculus is a branch of mathematics that deals with the behaviour of functions. It is used to determine the **gradient** of functions at any point without having to take a **tangent** to the **curve** at the point. This form of calculus is called *differential calculus* and is very closely related to the process described in the entry for **gradient** for a continuously bending curve. In fact, the **ratio** of Δy and Δx is replaced by the value of the gradient at the point; that is:

$$\Delta y/\Delta x \rightarrow dy/dx,$$

where dy/dx is the symbol for the gradient at the point. Since a curve is continuously bending, the gradient at a point on the curve is also a function of the **coordinates** of that point.

In general, the gradient of the tangent at any point (x, y) on the curve $y = ax^n$ is given by nax^{-1}. This general result is easily applied to the bx term in $y = ax^2 + bx + c$, since $bx = bx^1$ and on applying the above rule to find the gradient we get

$$2ax^1 + b.1.x^0 = 2ax + b.$$

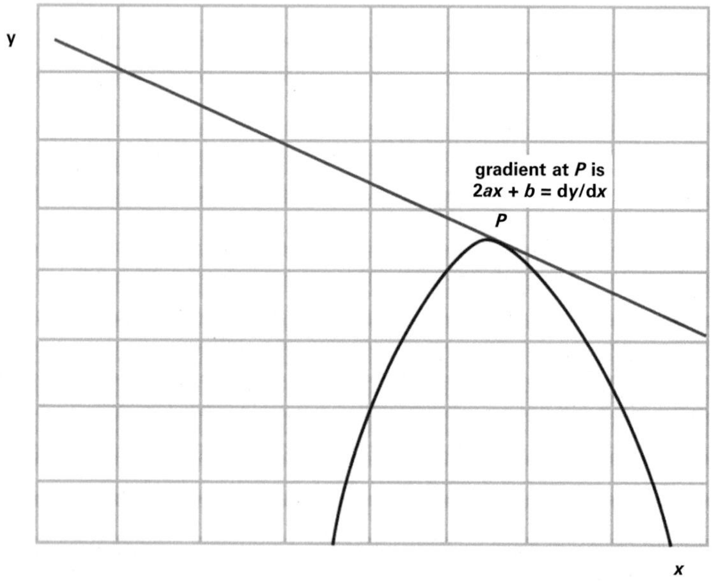

gradient at P is
$2ax + b = dy/dx$

P

gradient of a curve

For example, the gradient of the tangent at any point (x, y) on the curve whose equation is

$$y = ax^2 + bx + c$$

is given by the *gradient function* $= 2ax + b$. (See illustration on page 144.)

Another application of calculus is its use to calculate **areas** under parts of curves. This application of calculus is called *integral calculus* and in one formulation, is closely related to the area calculation made with **trapezia** in the entry **area under a curve**. The area under the curve is estimated with successively finer and finer trapezia. Each time the trapezia get finer, the estimate becomes better. The total area of all trapezia is the sum of all the areas. The shaded area in the figure opposite is being calculated in this manner. The sum of the trapezia is:

$$\sum \tfrac{1}{2}(h_i + h_{i+1})\Delta x$$

where h_i and Δx are respectively the vertical lengths and widths of the trapezia. The summation is taken from $i = 0$ to $i = n$, where n is the number of trapezia.

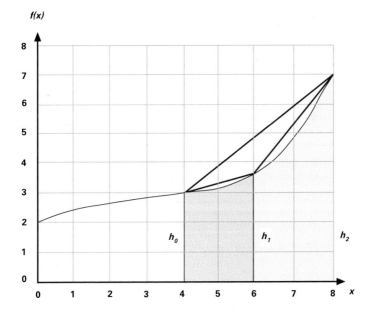

area under a curve

As we take finer and finer trapezia the sum becomes something called an *integral* and

$$\sum \tfrac{1}{2}(h_i + h_{i+1})\Delta x \to \int_{x(1)}^{x(2)} f(x)dx.$$

The long 'curly' S is a notation used to denote the **infinite** sum of all the little trapezium strips. Note that the expression in the integral is the function $f(x)$ because the strips have become so fine that we are not taking an approximation any more. The values of $x(1)$ and $x(2)$ are respectively the limits of the sum $x = 4$ and $x = 8$; that is, $x(1) = 4$ and $x(2) = 8$.

Appendix 2

Algebra
The ability to manipulate algebraic expressions is very important in any elementary mathematics course. This appendix deals with the rules of algebra for (1) rearranging an **equation**; (2) manipulation of algebraic **fractions**; and (3) manipulation of other **functions** (i.e. from **trigonometry** and **exponential** functions).

(1) *The balance method*
The operations **addition**, **subtraction**, **multiplication**, and **division**.
  When rearranging equations containing these operations, deal first with additions and subtractions. For example, let us make a the subject in the following equation:

$$(ab/c) + d - e = f$$

The equation reads: a multiplied by b and divided by c plus d minus e all equals f. Subtracting and adding d and e respectively from both sides of the equation leaves the equation unchanged but leaves:

$$ab/c = f - d + e,$$

which can now be multiplied by c and divided by b on both sides to leave:

$$a = (f - d + e)c/b$$

Note that when multiplying by c and dividing by b it was the whole expression in the **factor** $(f - d + e)$ that was affected.
  Brackets may be dealt with by making sure that anything multiplied or divided into the bracket is done so to *all* of its contents; that is, multiplication and division are **distributive**. For example we may *expand the bracket* as follows:

$$(f - d + e)c = (f - d + e) \times c$$
$$= f \times c - d \times c + e \times c,$$

$$(f - d + e) \div c = (f - d + e) \times (1/c)$$
$$= (f/c) - (d/c) + (e/c).$$

This converse is therefore also true; that is, a common factor may be taken out of an expression and a bracket used to show this (see **factor**).

This leads to the rule for multiplying two brackets, which may be summarized by the letters F.O.I.L. or First Outside Inside Last, referring to the positions of the objects to be multiplied in the two brackets. For example:

$$(a + b) \times (c + d) = a \times c + a \times d + b \times c + b \times d.$$

First: a and c are first letters appearing in the two brackets.
Outside: a and d appear on the outside boundaries of the expression.
Inside: b and c appear on the inside of the expression.
Last: b and d appear at the end of each bracket.

It is important to note that not all operations on a bracket are distributive. For example:

$$\sqrt{(x + y)} \neq \sqrt{x} + \sqrt{y},$$

$$(x + y)^2 \neq x^2 + y^2,$$

(2) The manipulation of algebraic fractions.

ADDITION: $a/b + c/d = (a \times d + c \times b)/(b \times d)$

SUBTRACTION: $a/b - c/d = (a \times d - c \times b)/(b \times d)$

Note that in both of these expressions a **common denominator** is used.

MULTIPLICATION: $a/b \times c/d = ac/bd$

DIVISION: $a/b \div c/d = a/b/c/d = (a/b) \times (d/c) = ad/bc$

The numbers a, d and b, c are sometimes said to be cross-multiplied.

Note that in the division the fraction c/d is inverted to make the division into a multiplication. This is done because

$$a/b/c/d = a/b \times 1/c/d$$

and $1/(c/d)$ is the **reciprocal** of c/d.

(3) Other functions

EXPONENTIAL FUNCTIONS AND INDICES:

The rules of indices for positive, negative, and fractional indices:

$$a^n = a \times a \times a \times a \times a \times a... \quad n \text{ times},$$

$$a^{-n} = 1/(a \times a \times a \times a \times a...) \quad n \text{ times},$$

$$a^{1/n} = \sqrt[n]{a}$$

The rules for exponential functions of x:

$$y = a^x \Rightarrow x = \log_a(y),$$

$$a^x a^y a^z = a^{(x+y+z)},$$

$$(a^x)^y = (a^y)^x = a^{xy} = a^{yx}.$$

The first of the above expressions comes from the definition of the **logarithm** of a number. Since x is the number that a has to be raised by to obtain y, x is said to be the logarithm of y to the **base** of a.

Worked example
Make x the subject of the following equation:

$$y = (a/b) \ exp(ikx)$$

The function $exp(ikx)$ (see **e**) is another way of writing e^{ikx}. The manipulation could go something like this:

$$y = (a/b) \ e^{ikx}$$
$$\Rightarrow by/a = e^{ikx}$$
$$\Rightarrow ikx = \log_e (by/a)$$
$$\Rightarrow x = (1/ik) \log_e (by/a).$$

TRIGONOMETRIC FUNCTIONS:
The trigonometric functions $\sin(x)$, $\cos(x)$, and $\tan(x)$ are also not distributive; that is,

$$\sin(x + y) \neq \sin(x) + \sin(y).$$

To manipulate trigonometric functions of compound angles such as $(x + y)$ one needs to refer to the compound angle formulae.

NOTATION:

$$y = \sin(x) \Rightarrow x = \sin^{-1}(y),$$

$$\sin(x) \times \sin(x) = (\sin(x))^2$$
$$= \sin^2(x),$$

$$\tan(x) = [\sin(x)]/[\cos(x)].$$

The notation $\sin^{-1}(x)$ is also sometimes written as $\arcsin(x)$. The notations in the first two lines above are also used for both $\cos(x)$ and $\tan(x)$, i.e.

$$y = \cos(x) \Rightarrow x = \cos-1(y),$$
$$\cos(x) \times \cos(x) = (\cos(x))^2$$
$$= \cos^2(x), \text{ etc.}$$

Worked example

Make x the subject of the following equation:

$$y = (a/b)\ sin(ikx).$$

The manipulation could go something like this:

$$y = (a/b)\ sin(ikx)$$
$$\Rightarrow by/a = sin(ikx)$$
$$\Rightarrow ikx = sin^{-1}(by/a)$$
$$\Rightarrow x = (1/ik)sin^{-1}(by/a).$$

Appendix 3

Coursework

Different examination boards have different requirements for coursework, but as a general guide, coursework is designed to test the student's ability to use and apply mathematics in everyday situations. Generally, the processes involved in the solution of a mathematical problem can be summarized by the following points:

(1) Break the problem down into simpler problems.
(2) Tabulate or neatly lay out the results of solving all the reduced problems.
(3) Look for a pattern in all these results.
(4) Use any pattern discovered to construct a mathematical expression. This expression can now be used as a predicting tool.
(5) Use the expression to predict more results which were not considered in the first part of the investigation. Verify that your expression leads to the same result as in practice.

These points are most easily illustrated by a worked example:

You work for a company that has just installed a new intercom system in its head offices. You are to investigate the various patterns of connections between rooms for this system. There are ten rooms to connect up.

It is obvious from the statement of this problem that we are not searching for a single solution as such. An investigation of the relationship between the number of rooms and the number of connections is what is required. Let us apply our five points above:
(1) To break the problem down, first consider a key for drawing diagrams of the intercom system. Let dots or **nodes** represent the rooms and **arcs** represent connections. First consider systems of 1, 2, 3, and 4 rooms rather than the full ten. Drawing systems of 3 and 4 rooms demonstrates that the problem quickly becomes complicated. See diagrams overleaf.
(2) What results are to be tabulated? To answer this question, let us consider what we are actually doing. We are looking for a relationship

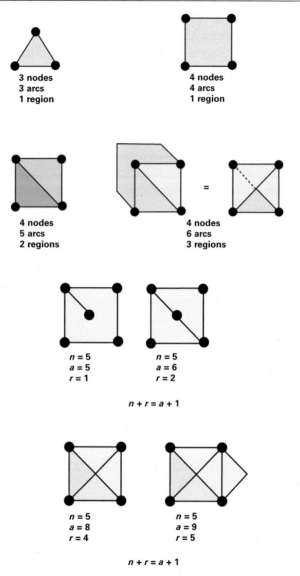

3 nodes
3 arcs
1 region

4 nodes
4 arcs
1 region

4 nodes
5 arcs
2 regions

=

4 nodes
6 arcs
3 regions

$n = 5$
$a = 5$
$r = 1$

$n = 5$
$a = 6$
$r = 2$

$$n + r = a + 1$$

$n = 5$
$a = 8$
$r = 4$

$n = 5$
$a = 9$
$r = 5$

$$n + r = a + 1$$

between the arcs (connections) and the nodes (rooms) so let us note the various networks according to their numbers of arcs and nodes. Looking at the various networks, we also notice that the number of regions cordoned off by the nodes and arcs changes.

(3) A pattern does emerge as follows:

Network with 3 nodes, 3 arcs, 1 region.

Network with 4 nodes, 4 arcs, 1 region.

Network with 4 nodes, 5 arcs, 2 regions.
Network with 4 nodes, 6 arcs, 3 regions.
(4) The emerging pattern is: number of nodes + number of regions = number of arcs + 1. If the number of nodes is n, the number of arcs is a and the number of regions is r then:

$$n + r = a + 1.$$

(5) For networks of five rooms we see the diagrams verify the above relationship. In fact this relationship is called **Euler's formula** after its discoverer the Swiss mathematician Euler (1707–1783). Since **vertices** could be thought of as nodes, Euler realized that with a small modification this formula could be applied to **polygons** and **polyhedra**; that is:

*number of faces + number of vertices
= number of edges + 1.*

Statistical Investigations

Some statistical investigations require the gathering and analysis of **data** to verify some **hypothesis**. The basic procedure below may help you plan such an endeavour:

MAKE A HYPOTHESIS

A hypothesis is a statement of an opinion about an issue. For example, the statement that 'most cars on the road are red' is an hypothesis. A **survey** can then be conducted to test this hypothesis; that is to find out whether or not the hypothesis is true.

DESIGN THE SURVEY AND DATA-COLLECTION SHEET

In designing your method of data collection, consider how you are going to conduct a fair test, i.e. note things that may have a negative effect on your survey. Whether or not your test is fair may also depend on control variables. For example, the fairness of a test of the above hypothesis might be affected if vehicles other than cars were included in the survey. The requirement that cars should be the only vehicles considered is a control variable.

COLLECT DATA

Tabulate your results to facilitate any later graphing of the data.

ANALYSE THE RESULTS

Draw any charts to demonstrate the validity or non-validity of your hypothesis. Finally comment on your results and suggest any improvements.

Writing a Report

When writing a report of your investigation, it is always useful to start by summarizing its aims. This will help you focus on the problem and also allow you to clearly state your intentions which will help anyone marking your work. A summary such as this is called an **abstract** and should appear at the beginning of your report after the title.

Your investigation will invariably involve making and verifying a hypothesis. Your report should reflect this by including sections corresponding to the various steps to making and testing a hypothesis. There are usually four steps to making and testing an hypothesis:

- Step 1: State the hypothesis based on preliminary observations.
- Step 2: Conduct an experiment to test the hypothesis. Collect relevant data and explain in this section of the report how this was done. Was it a fair test?
- Step 3: Analysis of the data. A full mathematical analysis should be included with an explanation in clear concise English.
- Step 4: Make a comparison of the results of the analysis and the original hypothesis. Report any similarities and anomalies and try to explain them.

A good conclusion is always necessary. It is the conclusion that is a final overview of the entire investigation. You should aim to cover the following points in any discussion:

(1) the making of the hypothesis from the preliminary observations,

(2) the decisions made leading to the design of the test,

(3) the results and any improvements that could be made to the gathering of data,

(4) the comparison of the results and the original hypothesis,

(5) the validity (or non-validity as the case may be) of the hypothesis.